Register No

THE NATIONAL SCHEDULE OF RATES

INFORMATION FOR PURCHASERS

The National Schedules are part of a complete option for undertaking term maintenance contracts. Purchasers rest in the knowledge that assistance is available from consultants who are all able, from experience in the field of measured term contracting, to assist users in the setting up and operation of this type of contract.

The National Schedules are fully resourced, updated and incorporate new items annually. They are providing an increasing number of owners and contractors with an efficient, consistent and accountable system for the commissioning of building repairs and maintenance works.

The cost of undertaking repairs and maintenance work can be estimated in advance of orders being placed. A further benefit to building owners and occupiers lies in the fact that standards of materials and workmanship can be firmly controlled and is likely to be of higher quality than the work valued on a daywork basis.

Apart from the printed version, the major National Schedules are available in a data format with integrated software. Samples of which can be downloaded from our website www.nsrm.co.uk

As a purchaser, you will receive the benefit of being able to share the experiences of other purchasers in operating measured term contracts by registering in the space provided below.

Those wishing to receive further detailed advice and assistance should contact the Administrator on: 01296 339966, or by fax 01296 338514 or email: nsr@nsrmanagement.co.uk.

It will help us to help you if you complete and return the information below:

REGISTRATION OF PURCHASERS - 08/09   No: _____   [from back of book]

Name: _____

Organisation: _____   Job title: _____

Address: _____
_____
_____

Telephone No: _____   Fax No: _____

email address: _____

How long have you been using NSR? _____

Named contractor using NSR on your behalf _____

Affix stamp here

**The Administrator
NSR Management
Pembroke Court
22-28 Cambridge Street
Aylesbury
HP20 1RS**

NATIONAL SCHEDULE of Rates 2008-2009

Published and Compiled Annually by:
NSR Management Ltd

General, Publications, Data Licence and Software available direct from:

NSR Management Ltd
Pembroke Court
22-28 Cambridge Street
Aylesbury HP20 1RS

Tel: 01296 339966
Fax: 01296 338514
www.nsrm.co.uk

Electrical Services in Buildings

Part 1

Introduction
Guidance to Users
Preliminaries/General Conditions
Materials and Workmanship Preambles
Measurement and Pricing Preambles

Part 2

Compiler's Notes
Schedules of Descriptions and Prices
Schedule of Basic Prices

ISBN 978-1-904739-38-8

Whilst all reasonable care has been taken in the compilation of the Electrical Services in Buildings Schedule the Co-authors and NSR Management will not be under any legal liability of any kind in respect of any mis-statement error or omission contained herein or for the reliance any person company or authority may place thereon.

Important

© All rights reserved. No part of any of the National Schedule of Rates may be reproduced stored in a retrieval system or transmitted in any form or by means electronic mechanical photocopying recording or otherwise without the prior consent of the Copyright owners.

This book is sold subject to the condition that it shall not, by way of trade or otherwise, be lent, resold, hired out or otherwise circulated without the publisher's prior written consent in any form of cover or binding other than that in which it is published and without a similar condition including this condition, being imposed upon the subsequent purchaser.

ELECTRICAL

PART 1

© NSR 2008 - 2009

Allow us to help you with the Property Maintenance Jigsaw

- Contract Advice
- QS Services
- Benchmark Services
- Condition Surveys
- DDA Surveys
- Management of Contracts

One STOP Shop

QS Services, Benchmarking, Condition Surveys

For over 12 years, NSR Management have been promoting and developing the National Schedule of Rates on behalf of the Co-authors, the Society of Construction and Quantity Surveyors and the Construction Confederation. During this time the portfolio of National Schedule of Rates have become synonymous with the management of Measured Term Contracts, and from 2006 NSR Management will compile the National Schedules.

Together with the appropriate contract conditions, these National Schedules allow customers to issue a series of works orders, confident that the charges for the work will be based on a pre-determined and agreed basis of measurement and pricing. Over the years the National Schedules have saved time and money for hundreds of organisations just like yours.

We like to think we listen carefully to our customers needs, and in response to their requests NSR Management now offer a comprehensive **One STOP Shop** consultancy service for those organisations involved in term contracting. Our team have over 70 years experience in the construction industry and we thought it time to let our customers have the benefit of this expertise.

Just look at some of the ways in which our comprehensive consultancy service can help you:

- Comprehensive advice on all aspects of Term Contracting
- Efficient preparation of Tender Documents to suit clients needs
- Sound Technical Advice
- Thorough Benchmarking and value for money assessments
- A wide range of other QS Services
- Smooth management of Measured Term Contracts
- Effective Stock Condition and DDA Surveys

NSR Management can offer flexible cost effective solutions to all of your property management needs. Combine this with the range of National Schedule of Rates, including our new web based schedules, and it is not hard to see why we lead the way...

Please contact us for more information

NSR
Management

NSR Management Limited
Pembroke Court
22-28 Cambridge Street
Aylesbury, Bucks HP20 1RS

tel: 01296 339966
fax: 01296 338514

nsr@nsrmanagement.co.uk

www.nsrm.co.uk

'Our business is your property'

General Index

Page

Part 1

Section A	: Introduction	1/1
Section B	: Contract Applications	1/5
Section C	: General Guidance to Users	1/7
Section D	: The Advisory Panel	1/13
Section E	: Tender Forms and Contract Documents	1/15
Section F	: Preliminaries/ General Conditions	1/19
Section G	: Material and Workmanship Preambles	1/39
Section H	: Measurement and Pricing Preambles	1/65

Part 2

Index	2/1
Summary of Amendments	2/5
Compiler's Notes	2/6
Schedule of Description and Prices	2/17
Basic Prices: Labour, Plant and Materials	BP/1

W T PARKER LIMITED

ELECTRICAL & MECHANICAL TERM CONTRACTORS TO DEFENCE ESTATES AND THE NHS

W.T. Parker Ltd
24-28 Moor Street
Burton on Trent
Staffordshire
DE14 3SX
Tel: 01283 542661
Fax: 01283 536189
www.wtparker.co.uk

SECTION A

INTRODUCTION

The National Schedule was first introduced by the Co-Authors - the Society of Chief Quantity Surveyors and the Construction Confederation in 1982 for use on maintenance and minor building work to meet the competition requirements of the Local Government Planning and Land Act 1980.

This schedule of rates for Electrical Services in Buildings has a similar format to that used for the National Schedule of Rates for Building Works, the prices shown against the description of work having been analysed and set down into their respective elements of labour, plant and materials. However the measurements of works, is defined in the item description, the measurement preambles and the Compiler's Notes.

The users attention is drawn to the materials prices used in this document. Wherever possible, the costs of materials are based on standard published price lists current at March/April 2008. Where these are not available, indicative quotations have been obtained for current trade prices. No discounts have been allowed against the materials costs and users are advised to check with their local suppliers to ascertain material costs and make due allowance in their tender percentage adjustment for such materials against those contained in the National Schedule.

The labour rates in this schedule include adjustments to the annual awards announced by the Construction Industry Joint Council and by the Joint Industrial Boards (JIB) for Plumbing & Mechanical Engineering Services and for the Electrical Contracting Industry. Detailed calculations of all allowances added to the basic rates of wages for craftsmen and labourers are given in the Compiler's Notes in Part 2 of the Schedule.

Whereas there is a requirement that all parties to the contract are required to comply with all relevant legislation current at the time of execution of the Works, users attention is particularly drawn to the 'Landfill tax', these regulations came into force on 1 October 1996, and were revised with effect from 1 April 1999, increasing the cost for disposal of active waste, the costs of which are to be included in the tendered percentage adjustment to schedule rates. (see page 1/9 for details of other percentage adjustment costs.)

The rates contained in this schedule do not include for any costs associated with 'The Construction (Design and Management) Regulations 1994'. Although it is envisaged that the majority of works undertaken will not require compliance with the CDM regulations, we would recommend that Clients appoint a CDM Co-ordinator and ensure that the tendered percentage adjustment includes for complying with the CDM Regulations and all as detailed in the approved Code of Practice – managing construction for Health and Safety, Construction (Design and Management) Regulations 1994.

This will ensure that a framework for complying with the CDM Regulations is in place and that the Contractor will be recompensed for his costs.

The rates contained in this Schedule do not include for any costs associated with working with asbestos. When carrying out work on asbestos based materials the contractor shall comply with the Control of Asbestos Regulations 2002 and appoint a licensed Specialist Subcontractor. Where minor work with asbestos is encountered and which can be carried out by a contractor who does not normally require an HSE licence as defined in the Approved Code of Practice and Guidance published by the Health & Safety Executive, the contractor may carry out the work as requested by the Contract Administrator at agreed rates or on a daywork basis [see page 1/26]

As an increasing number of property managers in both public and private sectors are forced to review their costs of building maintenance and repair they are considering the benefits to be gained from using a measured term contract, a comprehensive schedule of rates and a computerised management system such as is available through the National Schedule.

The value of the National Schedule to any public or private property owner or estate manager is best understood by considering how it may be used and the advantages it has over alternative methods of commissioning and paying for maintenance and minor works.

The use of the pre-priced and widely accepted National Schedule together with appropriate Contract Conditions allows the Client to issue a series of works orders, confident in the knowledge that the charges for the work when carried out will be made on a predetermined and agreed basis of measurement and pricing. Without the National Schedule the client either has to produce his own schedule, or invite a series of competitive quotations for the works before placing orders, or issue orders and subsequently be charged on a lump sum or daywork basis without prior knowledge of the end cost. In the latter case there may be no clearly defined contract conditions covering such matters as insurance, payment, or standards of workmanship and materials.

The published National Schedule data is produced and stored using computer software and is available under a licence to subscribers for use on their computer systems. For further details contact the Administrator on 01296 339966.

The Form of Tender is available from NSR Management Ltd. The JCT Measured Term Contract, together with Practice Note and Guide is available from NSR Management, RIBA, RICS etc. Standard Method of Measurement exception clauses are incorporated in the Preambles where appropriate.

The use of the National Schedule and the JCT Form of Measured Term Contract has been explained at conferences across the country organised by the Society of Construction and Quantity Surveyors (formerly the Society of Chief Quantity Surveyors), Local Government Associations and other organisations.

Other seminars covering the use of the National Schedule and the importance of Energy Efficient Design of buildings have taken place on a number of occasions in the South of England, promoted by Southern Electricity.

HISTORY

The need for a National Schedule for local authority maintenance and other building work arose from the change in requirements governing Local Government direct labour accounting brought about by the Local Government Planning and Land Act 1980.

In its Accounting Code of Practice for Direct Labour Organisations, the Chartered Institute of Public Finance and Accountancy recommend the use of schedules and this has been substantiated by the demands made by a large number of authorities. Statistics published by CIPFA show an increased use of schedules of rates in local authorities since 1981.

David Hannent, a Chief Quantity Surveyor, examined those schedules which were available at the beginning of 1982 and found that none fitted the needs of his authority. The Society of Chief Quantity Surveyors thereupon agreed to his proposal that they should adopt the development of a National Schedule. Discussions were held with a number of other local authority organisations and with the Construction Confederation.

The Construction Confederation had meantime produced a schedule and both the Construction Confederation and SCQS felt that their respective schedules had features worth adopting. It was agreed to merge the two documents and Contract Conditions, Form of Tender and Standard Minute Values were introduced at a later date, enabling users to invite tenders and analyse costs by means of a comprehensive range of documents.

A major change was introduced in 1990 to revise the descriptions, format and coding to comply with SMM7 and to introduce the JCT 1989 Conditions of Measured Term Contract with an updated Model Form of Tender. The Society of Chief Quantity Surveyors also published a revised Code of Procedure for Building Maintenance Works using the National Schedule in November 1997.

In 1991 the National Schedule published schedules of rates for Electrical and Mechanical services in Buildings and entered into a joint venture with the PSA on their updated schedule of rates for Landscape Management. A further joint venture led to the publication in 1994 of the National Roadworks Schedule of Rates. In 1995 the National Housing Maintenance Schedule was first published. This is a composite schedule of rates of approximately 400 items using the Building Works Schedule as a basis of pricing. In 2001, following customer feedback the Schedules for Electrical Services in Buildings and Mechanical Services in Building were comprehensively reviewed, updated and upgraded.

CHARACTERISTICS

The characteristics of the resultant National Schedule are:-

(i) Each rate is broken down into its component labour, materials and plant and may form the basis for a bonus measurement or material allocation.

(ii) Being produced and published by the Co-Authors as a national document, no extras apart from the annual cost of each edition will fall upon users.

(iii) It is published as a schedule of rates including other relevant information e.g. Tender Forms, Contract Conditions Preliminaries and Preambles, lists of basic prices of labour, plant and materials.

(iv) It complies with the Standard Method of Measurement 7th Edition (SMM7 amendment 3), except for minor variations specifically referred to in the Preambles section, so that measurement disputes are minimised.

(v) It may be used for works by private contractors as well as for DLO use.

(vi) It is computer based and, since it has a library of headings and descriptions by which it is produced, considerable further development is possible. The library has been structured so that new items may be inserted into their appropriate places allowing original items to retain their existing numbers.

(vii) It is comprehensive and although originally designed for public sector maintenance and repair work, it nevertheless may easily be adapted for other purposes. The schedules have been successfully used for fast-track rehabilitation contracts to reduce pre-contract periods.

(viii) It is acceptable to the industry as a whole and its flexibility allows continuous development.

(ix) An Advisory Panel comprises representatives of interested professional organisations and, being a fully constituted, independent body, provides the Co-Authors with advice and comment on any revisions of the National Schedule thought to be desirable.

(x) A Standard Minute Value Schedule is also available in a separate volume which gives an analysis of the labour content of each item.

(xi) Single trade schedules may also be produced from the available data. 'The Painting and Decorating Schedule of Rates' is currently available in this format.

(xii) It is sponsored by the Construction Confederation and the SCQS as the Co-Authors.

The Society of Construction and Quantity Surveyors and the Construction Confederation recognise the need for updating and improving any document of this kind and welcome suggestions and comments from subscribers. Technical queries and suggestions addressed to the Technical Secretary are referred where appropriate to the Advisory Panel and the Compilers and have led to the introduction of new items and other improvements in this and previous editions.

SECTION B

CONTRACT APPLICATIONS

The National Schedule has been priced from basic principles and is reviewed annually.

Each item in the Schedule is calculated using the appropriate materials prices, plant hire rates and all-in labour hourly rates as shown in the List of Basic Prices which is included in Part 2.

Tables showing the basis of the calculations of the all-in labour rates and a list of firms from whom material prices are obtained are included in the Compiler's Notes in Part 2.

The rates shown in the List of Basic Prices are nationally based and, by the use of percentage adjustments to the material, labour and plant values, the Schedule may be applied to any part of the United Kingdom. The rates do not however include profit and overheads and other costs which are noted in the contract conditions and preliminaries and should be allowed for as part of the tendered percentage adjustment to the rates.

As the National Schedule is divided into sections it may be used for tendering:-

(i)   As a complete schedule

(ii)   In groups of sections

(iii)   In individual sections

It may also be used with the following types of contract:

(i)   Fixed price

(ii)   Fluctuating - based upon changes in the National Schedule which is up-dated and re-published annually

(iii)   Fluctuating - other method, by agreement

It may be used to tender for:

(i)   All work in any geographical area

(ii)   Work in any particular type of building

(iii)   Work on an estate

(iv)   Term contracts

(v)   Lump sum contracts

With particular regard to its use by Direct Labour Organisations it must be emphasised that the National Schedule is aimed at providing a tender document based on current market prices. It may be used to provide cost targets after suitable adjustments. An Authority will be able to test the viability of its Direct Labour Organisation provided that a recent competitive tender, based on the National Schedule, has been obtained for work similar to that carried out by the DLO.

The National Schedule **with its adjusting percentage** gives the market value but the cost to users will be determined by their own accounting methods, usually by work-study-based bonus systems. In any case, it is necessary to link the National Schedule to the system and this may be possible, especially if the bonus system follows an established order. Any such set of bonus descriptions, when built up from or broken down to the National Schedule, will show an area of 'best fit'. The rogue items may then be priced separately to provide a direct local comparison.

SECTION C

GENERAL GUIDANCE TO USERS

Schedules of rates have been widely used for term contracts for maintenance and repair work for many years. The National Schedule is suitable for use with measured term contracts which include both reactive or day-to-day maintenance as well as pre-planned repair and improvement works. If the tender documents indicate that programmed cyclical maintenance is to be included it is probable that the tendered percentages will be lower as a result and it will also save the client the necessity of pre-pricing and inviting a number of individual tenders and contracts. Such works should not subsequently be taken out of the measured term contract to achieve more competitive bids for remaining reactive maintenance and repair work.

Term Contracts

Where an employer wishes to engage a contractor to carry out ad hoc maintenance and repair work and also requires competitive quotations for the work, experience has shown that term contracts based on a priced schedule of rates may be suitable. However the following factors should be noted:-

(a) Within certain broad parameters, such as overall value, location and time, the term contract is for an unspecified amount of work at an unspecified location at an unspecified time.

(b) This type of contract is intended to cope with large numbers of small orders placed over the whole term of the contract and this allows the statistical "swings and roundabouts" to operate to reduce the overall pricing risk to reasonable proportions.

(c) In arriving at his percentage on or off the schedule of rates the contractor will have to price what he thinks will be a representative selection of items of work to be done. The tender invitation should include a clear description of the types of work to be done and the pricing restricted to those sections of the schedule covering the intended work. For example it would be wrong to invite tenders for work which is described as predominantly electrical work and issue substantial orders for fire protection. Such requirements for unrepresentative work could seriously affect the pricing calculations and could not be held to come within the normal "swings and roundabouts" of the system.

(d) Both contractors and employers need experience with a schedule system before prices can be stabilised. Users are therefore reminded that they should look upon the introduction of a schedule of rates as a long term project and ensure that they have adequate technical support services in their organisation.

(e) Contract conditions are specifically designed for this type of work.

Single Projects of Refurbishment or Modernisation Work

Where a single substantial package of work can be identified it might be advantageous to tender separately on the basis of the schedule. In these circumstances more specific information may be given about scope, timing and location of the work and the contractor will price with greater confidence. For example, the kind of project envisaged for this approach might be an estate of houses or a school or college which are to be modernised and redecorated. The description of the work would be backed up by drawings, specifications, schedules and project preliminaries.

The contract conditions for this type of project could be carried out under the JCT 98 or 2005 Forms with approximate quantities where the quantities are priced at National Schedule rates to which tenderers will apply a percentage adjustment.

Where it is not possible to measure in sufficient detail and the Measured Term Contract is used, it is nevertheless recommended that some pre-measurement or assessment is undertaken to evaluate the scope of the contract to attract the best tenders.

Single New Work Projects

As explained above the more specific the description of the work, the less the risk in pricing for both contractor and employer. It is possible to fully describe new work through drawings, specifications and bills of quantities, which may be priced relative to the National Schedule if desired.

Selection of Contractor

The Form of Tender for use with Measured Term Contracts let on the National Schedule provides for tenderers to insert a percentage adjustment to the rates listed in the National Schedule for work to be measured and valued in accordance with the contract and for other percentage adjustments to the prime cost of labour, materials, plant and subcontractors for work carried out or valued under the contract as daywork.

It is recommended that the client informs the tenderers of the value of dayworks that will be assessed in the selection process as different combinations may give different results.

In the new MTC contract the contractors have to submit a schedule of hourly charges. The invoice price of non labour items are then subject to percentage addition for overheads and profit. These percentage additions can then be assessed as below.

In the table, contractor C is the lowest price by virtue of the more competitive dayworks percentages, after allowing for the assessed values of measured work and dayworks, as is seen from the calculation of anticipated final account costs.

TABLE

Contractor	PERCENTAGES IN TENDER			
	Percentage Adjustment	DAYWORK		
		Material	Plant	Subcontract
A	+10	20	15	15
B	+11	15	10	10
C	+12	5	5	5

ASSESSMENT

If, for example, the client estimates that in every £100,000 of the final account £85,000 will be valued at schedule rates or pro rata, and that there will be; £5,000 labour; £4,000 materials; £3,000 plant and £8,000 sub-contract values in dayworks, the estimated final account values of the three tenders will be:-

Net Value of Work			Final Account Values		
		£	Contr. A £	Contr. B £	Contr. C £
Schedule		80,000	88,000	88,800	89,600
Dayworks -	Labour	5,000	5,000	5,000	5,000
	Materials	4,000	4,800	4,600	4,200
	Plant	3,000	3,450	3,300	3,150
	Sub-Contr	8,000	9,200	8,800	8,400
Total		100,000	110,450	110,500	110,350

This Assessment assumes a contract where little overtime will be instructed by the Contract Administrator. Where a contractor is expected to carry out a significant part of the contract outside normal working hours an allowance in respect of overheads and profit on non profit overtime rates [Percentage adjustment of Clause 5.7 of the JCT Measured Term Contract] should be incorporated into the tender assessment.

Another factor recommended to clients who order a regular amount of this work is to maintain a separate select list of contractors for schedules. In this way the contractor may establish a workforce with knowledge and experience of schedules, have surveyors and managers who understand the system and the client's requirements, and thus provide the most competitive price.

Certainly, contractors are now available who specialise in National Schedule work and these should be sought. The other advantage in having a specialist list is that it gives the inevitable "swings and roundabouts" a chance to work.

It is significant that tenders are now stabilising as experience of methods and pricing levels is obtained. Encouragement and stimulation given to 'schedule keen' contractors will provide an increasing pool of knowledge which will benefit all parties and is certain to reduce costs and prices in the long term.

Tendered Percentage Adjustment

The tendered percentage is deemed to include but not limited to:-

Contractor's overheads and profit
Travelling time where applicable
Permit to work
Setting out and checking for dimensions
Off-loading, manhandling and transport and delivery of materials and plant
Square cutting and waste where different from allowance included in rates
All necessary security precautions
Protection, drying and cleaning – see item in Section F on page 1/27
Compliance with Regulations
Safety Health and Welfare regulations and policies
CDM Regulations
Landfill tax
Control of noise
Any site establishment costs necessary including temporary telephones
Rubbish removal including tipping
Management and staff
Disbursements arising from the employment of work people
Service and facilities
Hand tools and mechanical plant where not included in Part 2
Temporary works
Scaffolding –   temporary staging up to 1.5 metres and ladders for access to buildings of not more than 2 storeys in height
Client training
Adjustment to standard specification requirements notified to tenderers and/or variations for local conditions
Records and drawings
Identification and labelling of services
Builders work information (rates are provided for chases and holes etc)
Work in conjunction with ceiling installations
Painting and priming of services
General bonding and earthing
Inspection testing and commissioning of the works (see also references to testing on pages 1/33, 1/40 and 2/14)
Complying with preliminaries particulars, conditions of contract and preambles
Trade discount adjustments

Specialist Goods and Work not specified within the Schedule

There will be some items of work which are not suitable for inclusion in a standard schedule of rates because the circumstances on which they are required and consequently the cost will vary considerably e.g. asbestos removal etc. There are other non-standard items which are required only rarely. Sometimes it will be better for a client to have separate contracts for such works, however where these works are included as non-standard items on orders given to the Term Contractor the following procedures are to be adopted.

The Contractor is entitled to a 5% cash discount on the nett cost of specialist goods and a 2.5% cash discount on the nett cost of specialist work. Appropriate adjustments to achieve these percentages will be made when compiling the final account.

The nett cost of specialist goods and work will NOT be subject to any tendered percentage (omission or addition) but the Contractor is entitled to a percentage addition for overheads, profit and general attendance on the cost of specialist goods and work **after** the adjustment to provide cash discount described above.

Specialist attendance shall be measured separately in accordance with the provisions of the Standard Method of Measurement.

The nett cost is defined as the cost charged to the contractor after deduction of all trade or other discounts except for cash discounts referred to above.

NOTE: Where items containing Prime Cost sums have an identical fix only item then the materials used for the fix only item will NOT be treated as specialist goods but adjusted for waste factors, cash discounts and the tender percentage (omission or addition) as for Prime Cost items.

Prime Cost Items specified within the Schedule

All Prime Cost Items will be adjusted in the final account. The Contractor is entitled to a 5% cash discount on the nett cost of goods supplied under instruction against a Prime Cost Item specified within the Schedule. Appropriate adjustment to achieve this percentage will be made when compiling the final account.

The effect of the adjustment of the cost of the Prime Cost Item (after the adjustment to provide cash discount described above) will be subject to any tendered percentage (omission or addition).

The nett cost is defined as the cost charged to the contractor after deduction of all trade or other discounts except for cash discounts referred to above.

No waste factors have been allowed relative to prime cost items.

Overtime Payments

Overtime payments are to be paid in accordance with Clause 5.7 of the Standard Form of Measured Term Contract which refers to Working Rule 4, published by Construction Industry Joint Council, which states when overtime payments should commence.

The rates for overtime should be paid as follows:-

Mondays - Fridays

First four hours	-	Time and a half
After first four hours until normal starting time next morning	-	Double time

Saturdays

First four hours	-	Time and a half
After first four hours	-	Double time

Sundays

All hours	-	Double time

Notwithstanding the requirements of the Standard Form an easy administrative way of dealing with out of hours work is to adopt the measures set out in the following example.

Work carried out being valued using Schedule Rates:-

Example

	£	£
V 8568/700		15.48
Add tendered percentage adjustment say 10%		<u>1.55</u>
		17.03
Add non productive overtime (time and a half)		
50% of labour £10.15 x 50%	5.07	
Add tender percentage for overtime adjustment say 15%	0.76 +	<u>5.83</u>
	£	<u>22.86</u>

Work carried out being valued using Daywork Rates:-

Example

	£	£
3 hours @ £25.00		75.00
Add non productive overtime (double time)		
100% x 3 hours @ £25.00	75.00	
Add tender percentage for overtime adjustment say 50%	37.50 +	<u>112.50</u>
	£	<u>187.50</u>

Procedures

Utilising the documents requires the client to think carefully about procedures and staffing, especially in the context of term contracts. This can only be done at local level but is preferably defined in the tender documentation. No national system can be devised to satisfy all requirements, and the National Schedule is not a panacea for this headache. The matters of progress of job tickets, priorities, emergency work, valuations, dayworks, supervision and access are all specific to the individual client. Clauses resolving these items may be inserted in the preliminaries or within any relevant explanatory procedure document. Suggested headings for preliminaries clauses are given elsewhere (see Section F).

Appendix A of Section F contains a list of supplementary clauses, which a client may wish to include in contract documentation dependent on particular circumstances. Advice on the incorporation of these clauses and other more particular advice on the contents of Conditions of Contracts and Preliminaries is available from NSR Management on a consultancy basis.

User Feedback

In order to continuously update and improve the Schedules, feedback is required from users and clients alike. Please return the User Response Form issued with this documentation to the Administrator, NSR Management, Pembroke Court, 22-28 Cambridge Street, Aylesbury, HP20 1RS.

SECTION D

THE ADVISORY PANEL

The panel has been established as an independent body providing a forum for discussion and feedback to the Co-Authors.

The Panel's role is seen to be one which provides users with an independent committee capable of directing the needs of the user. The Co-Authors are committed to providing the Panel with the facilities to act as such.

Members of the Advisory Panel represent the following bodies concerned with the execution of construction and maintenance work:-

1. British Constructional Steelwork Association Ltd
2. British Gas
3. Construction Confederation
4. Chartered Institute of Public Finance and Accountancy
5. Chief Building Surveyors Society
6. Society of Chief Architects
7. Society of Construction and Quantity Surveyors
8. Society of Electrical and Mechanical Engineers in Local Government

SECTION E

TENDER FORMS AND CONTRACT DOCUMENTS

The Form of Tender and the Contract Conditions are available as separate publications. The Forms of Tender are available from NSR Management Ltd and the Contract Conditions are published by RIBA Publications and available through their usual outlets and NSR Management Ltd.

It is recognised that some alterations may be made in applying these forms to particular contractual requirements, but as printed they provide a model format suitable for most requirements.

In view of the variety of possible combinations of tender the following must be considered as a guide. It is possible to tabulate the requirements of a tender thus:-

(a) **Which sections of the National Schedule will apply?**

(b) **Are dayworks to be appended?**

(c) **How is scaffolding to be paid for?**

(d) **How is travelling to be paid for?**

(e) **How is plant to be paid for?**

(f) **How are the CDM regulations to be paid for?**

(g) **What is the definition of the area of operation of the contract?**

(h) **How is emergency work to be paid for?**

(i) **What is duration of the contract?**

(j) **Is the contract fixed price or fluctuating?**

(k) **How is the work to be measured?**

(l) **What type of work is to be executed?**

(m) **What preliminaries and preambles apply?**

(n) **What comprises the Contract Documents for tendering?**

NOTES

(a) **Which sections will apply?** - Rather than have a tender which has a different percentage against each section it is advised that a single percentage is applied.

(b) **Dayworks** - Daywork should be calculated in accordance with JCT MTC clause 5.4.

(c) **Scaffolding** - Further details on this may be found in the Preliminaries in Section F; rates for scaffolding are included in Part 2 (Schedule of Descriptions and Prices).

The rates contained within this schedule, together with the Percentage adjustment include for working within high rise buildings and for working from scaffolding up to a height of 10 metres above ground level.

Where external operations are to be carried out from scaffolding or other methods of access above 10 metres the tender documents should give specific information on how such orders are to be executed and, where not deemed included in the Percentage adjustment, allow for an additional percentage adjustment to the rates depending on the height eg:

    10 - 15 metres 2.5%
    15 - 20 metres 3%
    20 - 25 metres 3.5%
    25 - 30 metres 4%

(d) **Travelling** - The cost of this item should be included in the tender percentage adjustment. Where the properties to be maintained are spread over a wide area this could be a significant overhead cost and relevant information should be given to tenderers. Some contracts provide for adjustments to tendered percentage to allow for different distances travelled to jobs.

(e) **Plant** - The cost of plant such as compressors, kango hammers, drills etc. is included where appropriate and identified in the break down of the rates shown in Part 2. If the client intends to make some plant available to the contractor it should be clearly identified in the preliminaries giving details of the terms under which it may be used.

(f) **CDM Regulations** - The Client should appoint a CDM Co-ordinator and ensure the Principal Contractor prepares a construction phase plan. The cost to the Contractor in complying with the CDM Regulations should then be included in the tender percentage.

(g) **Definition of area** - The tenderer must be advised of the location and conditions affecting his work. For example, an 'area' could be an estate, a building or a geographical area.

(h) **Emergency Work** - During normal working hours emergency work may be paid for under the provisions of clause 5.8 of the JCT Measured Term Contract. Such work outside normal working hours may be paid at daywork rates and may involve a minimum call-out charge. The tender documents should give specific information on how such orders will be executed and paid for and what levels of priority will be used. Priorities may be listed as footnote (7) to Contract Appendix item 6.

Further requirements for the provision of emergency services may be given as follows:

Emergency Services

The contractor must provide and maintain an out-of- hours emergency service to avoid danger to the health and safety of residents and the public or serious damage to buildings and other structures. The service must be provided 24 hours each day including at weekends, and during holiday periods.

The names, addresses and telephone numbers of suitable competent persons who may be contacted outside working hours shall be provided by the contractor. The telephone must be manned and not an answering machine or answering service or mobile service.

(i) **Duration** - The exact dates for commencement and completion of the contract must be defined.

(j) **Fixed price or fluctuating** - Alternative methods may be adopted.

For fixed price work it is only necessary to identify the rates which will apply by reference to a particular edition of the schedule.

For fluctuating contracts, which are expected to be for periods over one year duration, it is recommended that a fixed date be specified when the rates will be varied, i.e. all work executed after 1st August will be priced at that schedule rate until 31st July the following year.

It may be more convenient to apply a change date to the date the work is measured but this must be agreed by the user.

(k) **How is the work to be measured?** - The work may be measured by the Contract Administrator; alternatively the contract provides for the Contractor to carry out the appropriate measurement for subsequent verification of the account by the Contract Administrator. Where the client intends that the contractor shall measure the works in whole or in part this is to be stated in item 9 of the Appendix to the contract.

(l) **What type of work?** - The National Schedule does not define what information must be given at tender stage about the type of work involved and the range expected, eg. "all repairs to electrical work". Further information may be given as follows:-

"The scope of work comprised in any order or in the whole contract cannot be pre-determined and no undertaking is given regarding continuity or overall value of the work and the contractor must allow in his tender for all intermittent or abnormal workloads.

As an indication of the anticipated scope of the work the following information is given as a guide:

No. of properties..............
No. of orders per week...............
Average value of order...............

Average classification of work:-

Heating installation	....%
Power installation	....%

Work in other trades will also be required."

(m) **Preliminaries and Preambles** - Model Preliminaries and Standard Preambles are included as part of the Schedule. It is important that the Contract Administrator should define the conditions and the character of work as carefully as possible at tender stage and the circumstances in which it is executed.

(n) **The Contract Documents for Tendering** - Contractors prepared to tender will be expected to subscribe to the National Schedule of Rates. Letters of invitation from the Client will then be required which will incorporate the tender documentation set out below.

Recommended Tender Documentation from the Client:

This would include:

    (i) **Letter of invitation**

    (ii) **Tender Form**

This must note the Contract Conditions if different from those included in the schedule of rates.

    (iii) **Schedule of Works**

The Model Preliminaries allow for Description of Works, however, as much detail as possible of the actual work to be done should be provided and this is best achieved by way of an accurate Schedule of Works. The Schedule of Works may be the job ticket(s) issued during the contract, a prepared Schedule of the Works actually required, or an overall definition of the work to be done when and if circumstances arise. As much information as possible at tender stage will assist in obtaining a relevant and competitive tender.

(iv) **Preliminaries & Preambles**

The relation of all documents to one another is most important and any adaptation of Model Preliminaries and Standard Preambles should be done with care to ensure that they are compatible. The appendix to the contract must be completed by the client.

(v) **Drawings**

Any drawings that are available will assist the tenderer.

(vi) **Other information**

All attached correspondence and information should relate to the main documents and indicate the exact content of the Tender and Contract Documentation. The date and time for receipt of Tenders and a plain addressed envelope for their return to the client should be provided.

Further information on the procurement and administration of term contracts is contained in a Code of Procedure For Building Maintenance Works published jointly by the Society of Construction and Quantity Surveyors and NSR Management.

SECTION F

PRELIMINARIES/GENERAL CONDITIONS

The following information is given to provide general and particular information relating to the proposed contract which could affect the tender and to assist in the completion of the appendix to the JCT Measured Term Contract Conditions pages 5 - 10.

Appendix A contains a list of supplementary clauses which may be applicable on a particular contract dependant on circumstances.

CONTENTS

	Page
PROJECT PARTICULARS	1/20
DRAWINGS	1/20
THE SITE/EXISTING BUILDINGS	1/20
DESCRIPTION OF THE WORK	1/20
THE CONTRACT/SUB CONTRACT	1/21
EMPLOYERS REQUIREMENTS	1/21
: Tendering/Sub letting/Supply	1/21
: Provision, content and use of documents	1/22
: Management of the Works	1/22
: Quality Standards/Control	1/22
: Security/Safety/Protection	1/24
: Specific limitations on method/sequence/timing	1/28
: Facilities/Temporary work/Services	1/30
: Operation/Maintenance of the finished works	1/32
CONTRACTOR'S GENERAL COST ITEMS	1/32
WORK/MATERIALS BY THE EMPLOYER	1/33
NOMINATED SUB CONTRACTORS AND SUPPLIERS	1/33
WORK BY STATUTORY AUTHORITIES	1/33
APPENDIX A – LIST OF SUPPLEMENTARY CLAUSES	1/35
APPENDIX B – GENERAL ADVICE ON TUPE	1/37

PROJECT PARTICULARS

Employer : Name and address of Employer to be given

Location : The site(s) is/are situated *

Access : Provision for access will be arranged by the Contract Administrator in accordance with clause 3.4 of the Contract Conditions, access to the site(s) is by .... *

The Contractor is deemed to have visited the site(s) and ascertained the means of access and any limitations thereto: no claims for additional costs caused by access difficulties arising from the execution of the works and which were identified at the date of tender will be allowed.

The Contractor is to agree access to the Site(s) with the Contract Administrator before commencement of the works.

DRAWINGS

The following drawings will form part of this contract:

Drawing No	Description
....	....
....	....

Further contract drawings may be issued during the term of the contract.

THE SITE/EXISTING BUILDINGS

Site boundaries: Where works are to be carried out on open sites the boundaries and other details must be provided with the order.

Existing buildings: Work to existing buildings should be clearly defined by room or dwelling number or drawing as necessary.

Existing mains or services: Contractors attention should be drawn to services in the vicinity of proposed works and any limitations regarding their use or protection should be given.

DESCRIPTION OF THE WORK

General Description: The works covered by this contract and these schedules of rates comprise ....

.... *

* Contract Administrator to complete/delete as applicable

THE CONTRACT/SUB-CONTRACT

Conditions of Contract: The Form of Contract will be the JCT Measured Term Contract 2006 Edition with amendments unless otherwise stated. NOTE – This contract makes no provision for Nominated Suppliers or Sub-Contractors.

(Where an alternative form of contract is selected a schedule of the clause headings should be given).

Special conditions or amendments to standard conditions should be stated.

All insertions to the contract appendix are to be made including employers insurance responsibility.

Performance Guarantee Bond

Note   This may not be required for measured term contracts where completion of works ordered occurs regularly throughout the term of the contract. Similarly a retention percentage of stage payments is not needed for the same reason.

EMPLOYERS REQUIREMENTS

Tendering/Sub-letting/Supply:

Details to be given of any restrictions or requirements e.g.

MATERIALS to be supplied by client.

TENDERING to be in accordance with the principles of the Code of Procedure for Single Stage Selective Tendering 1996 (Alternative 1 or 2).

AMENDMENTS: No amendments to be made to the schedules or other contract documents by tenderers. Tenders containing amendments or qualifications may be rejected.

COSTS TO BE INCLUDED: The contractor is to allow for costs of fulfilling all liabilities and obligations referred to in the preliminaries, preambles and other tender documents as part of his tendered percentages.

FIXED PRICE: All works carried out under the contract shall be valued in accordance with the published edition of the schedule of rates named in the tender and contract subject to the tendered percentage adjustment *

or

FLUCTUATING PRICE: All works carried out under the contract shall be valued in accordance with the edition of the schedule of rates current at the date the work is carried out subject to the tendered percentage adjustment.*

(Note this does not apply to dayworks)

TRANSFER OF UNDERTAKINGS (PROTECTION OF EMPLOYMENT) REGULATIONS 1981 AND THE ACQUIRED RIGHTS DIRECTIVE 1977: The requirements of these regulations may or may not* apply to this contract and professional advice should be sought. General advice is contained in Appendix B.

---

* Contract Administrator to complete/delete as applicable.

Provision, content and use of documents

All tender documents may be inspected by appointment during normal office hours at the employer's offices situated at*

....

....

Tel......................................

Tenderers are recommended to obtain their own copy of the National Schedule which includes the Conditions of Contract, Preliminaries, Preambles together with the schedule of descriptions and prices.

SCHEDULE COMPLIES WITH SMM7: the Schedule of Rates has been prepared generally in accordance with the Standard Method of Measurement of Building Works, 7th Edition (amendment 3), subject to amplifications inherent in any of the descriptions therein. Items or methods of measurement contrary to the Standard Method of Measurement of Building Works are stated in the Preambles relating to the Schedule of Rates.

PREAMBLES QUALIFY DESCRIPTIONS: items in the Preambles relating to the Schedules of Rates are deemed to qualify and to be part of every description to which they refer.

RECIPROCAL USE OF RATES; rates may be used reciprocally between sections of the Schedules of Rates in the settlement of accounts.

Management of the Works

PERSON IN CHARGE: During the carrying out of the works, the Contractor is to keep on the works a competent person in charge who shall be empowered to receive and act upon any instructions given by the contract administrator or his representative.

Quality Standards/Control

QUALITY CONTROL: the contractor is to establish and maintain procedures to ensure that the works including the work of all subcontractors, comply with specified requirements. Maintain full records, keep copies for inspection by the Contract Administrator and submit copies on request. The records must include:-

a) The nature and dates of inspections, tests and approval
b) The nature and extent of any nonconforming work found
c) Details of any corrective action

SETTING OUT: the Contractor is to take the dimensions from existing premises and check with dimensions given on the drawings. Allow for setting out the works and providing all instruments and attendance required for checking by the Contract Administrator.

MATERIALS, LABOUR AND PLANT: provide all materials, labour and plant and all carriage, freightage, implements, tools and whatever else may be required for the proper and efficient execution and completion of the works. All materials are to be new unless otherwise specified.

OFF-LOADING AND MANHANDLING: the Contractor shall allow for all off-loading and manhandling into position including placing into and removing from temporary site storage prior to final positioning, all general materials, plant and items of equipment.

* Contract Administrator to complete/delete as applicable.

SAMPLES AND STANDARDS OF MATERIALS: the Contractor shall allow for obtaining samples of materials as required by the Contract Administrator. Such samples to be approved by the Contract Administrator before use or application in the works. All material subsequently used in the works is to be of equal quality in all respects to the approved sample.

MANUFACTURER'S RECOMMENDATIONS: means the manufacturer's recommendations or instructions, printed or in writing current at the date of tender.

COMPLIANCE WITH REGULATIONS: the Contractor shall allow in his tendered percentage 'A' to the schedule of rates for and ensure that the works and components thereof comply with all recommendations, requirements and current editions of the following:-

a) British Standard Specifications

b) British Standard Codes of Practice

c) The Factories Act

d) The Health and Safety at Work Act

e) The Gas Act and Regulations

f) Electricity at Work Act and Regulations

g) The Regulations for Mechanical Installations issued by the Heating, Ventilation and Air Conditioning Association

h) The Regulations for Electrical Installations issued by the Institution of Electrical Engineers

i) The Electrical Supply Regulations

j) The Rules and Regulations of the Local Electricity, Gas and Water Authorities

k) Fire Precaution Act and Regulations and the requirements of the Local Fire Authorities

l) Portable Appliance Regulations

m) Regulations relating to the control of Asbestos and Legionnaires Disease

n) Regulations relating to Waste Management

PROPRIETARY NAMES: the phrase 'or other approved' is to be deemed included whenever products are specified by proprietary name.

INCOME AND CORPORATION TAXES ACT 1988 or any statutory amendment or modification thereof: the attention of the Contractor is drawn to this Act which replaces the Finance (NO.2) ACT 1975. The provisions of this Act are explained in the Board of Inland Revenue pamphlets.

The Contractor is also reminded that it is his duty and responsibility to satisfy himself that all Sub-Contractors are approved by the Contract Administrator and hold an appropriate Sub-Contractor's Certificate from the Inland Revenue.

DIMENSIONS: dimensions stated or figured dimensions on the drawings are to be adhered to. Any discrepancy between the drawings is to be brought to the notice of the Contract Administrator for clarification and instruction.

GENERALLY: all standards referred to within these documents shall be held to be the latest edition published at the date of tender. A reference to any Act of Parliament, Regulation, Code of Procedure or the like shall include a reference to any amendment or re-enactment of same.

CO-ORDINATION OF ENGINEERING SERVICES: The site organisation staff must include at least one person with appropriate knowledge and experience of mechanical and electrical engineering services to ensure compatibility between engineering services, one with another and each in relation to the works generally.

REINSTATEMENT: the Contractor shall make good and re-instate in working order any existing security system switched off, damaged or otherwise rendered inoperable by the works.

Security/Safety/Protection

SECURITY PRECAUTIONS: the Contractor is to allow for any security precautions that may become necessary in relation to the adjoining properties during the course of the works and is to allow for adequate measures to prevent access from scaffolding or similar means.

The Contractor shall issue ID badges to all employees who are engaged on the Works and these badges are to be displayed at all times. Contractor employees shall wear suitable clothing to identify them as employees of the Contractor.

SAFEGUARDING THE WORKS, MATERIALS AND PLANT AGAINST DAMAGE AND THEFT: the Contractor shall provide for all necessary watching and lighting and care of the whole works from weather, theft or other damage. All materials on site shall be protected from damage or loss.

STABILITY: the contractor shall accept responsibility for the stability and structural integrity of the works and support as necessary. Prevent overloading.

TRESPASS AND NUISANCE: all reasonable means shall be used to avoid inconveniencing adjoining owners and occupiers. No workman employed on the works shall be allowed to trespass upon adjoining properties. If the execution of the Works requires that workmen must enter upon adjoining property, the necessary permission shall be first obtained by the Contractor.

The Contractor shall not obstruct any public way or otherwise permit to be done anything which may amount to a nuisance or annoyance, and shall not interfere with any right of way or light to adjoining property.

TRAFFIC AND POLICE REGULATIONS: all traffic and police regulations particularly relating to unloading and loading of vehicles must be complied with and all permits properly obtained in due time for the works.

CONTROL OF NOISE: ensure that all measures are taken to control noise levels in accordance with the Noise at Work Regulations 1989, the control of Pollution Act 1974, the control of Noise (Code of Practice for Construction Sites) Order 1975 and BS 5228 including complying with DOE advisory leaflet 72 – noise control on building sites. Compressors, percussion tools and vehicles shall be fitted with effective silencers of a type recommended by the manufacturers of the compressors, tools or vehicles. The use of pneumatic drills and other noisy appliances must have the Contract Administrator's consent. The use of radio or other audio equipment will not be permitted.

POLLUTION: take all reasonable precautions to prevent pollution of the works and the general environment. If pollution occurs inform the appropriate Authorities and the Contract Administrator without delay and provide them with all relevant information.

NUISANCE: take all necessary precautions to prevent nuisance from smoke, dust, rubbish, vermin and other causes.

SAFETY HEALTH AND WELFARE: allow for complying with all Safety, Health and Welfare Regulations appertaining to all workpeople on site including those employed by sub-contractors and professional advisers, including but not limited to the following and to include any future amendments or re-enactments thereto:-

    i)   The Construction (Design and Management) Regulations 1994

    ii)   The Construction Regulations 1961, 1966 and 1996

iii) The Factories Act 1961

iv) The Offices, Shops and Railway Premises Act 1963

v) Work Equipment Regulations 1998

vi) Management of Health and Safety at Work Regulations 1999

vii) The Health and Safety at Work, etc Act 1974

viii) Manual Handling Operations Regulations 1992

ix) Workplace (Health Safety and Welfare) Regulations 1992

x) Personal Protective Equipment at Work Regulations 1992

xi) Display Screen Equipment Regulations 1992

xii) Special Waste Regulations 1996

xiii) Control of Asbestos at Work Regulations 2002

xiv) Construction (Head protection) Regulations 1989

xv) Construction (Health and Safety and Welfare) Regulations 1996

xvi) Control of Substances Hazardous to Health Regulations 2002

xvii) Environmental Protection Act 1990

xviii) Reporting of Injuries, Diseases and Dangerous Occurrences Regulations 1995

xix) Waste Management Licensing Regulations 1994

xx) Gas Safety (Installation and Use) Regulations 1994

xxi) Fire Precautions (Workplace) Regulations 1997

xxii) Lifting Operations and Lifting Equipment Regulations 1998

xxiii) Provision and Use of Work Equipment Regulations 1998

xxiv) The Secure Tenancies (Right to Repair) Regulations 1985

xxv) The Equal Pay Act 1970

xxvi) The Employment Protection (Consolidation) Act 1994

xxvii) The Enterprise Act 2002

xxviii) The Sex Discrimination Act 1975 and Disability and Discrimination Act

xxix) The Race Relations Act 1976 and the Race Relations Amendment Act 2000

xxx) Health and Safety (Safety, Signs and Signals) and [Consultation with Employees] Regulations 1996

xxxi) Management of Health and Safety at Work and Fire Precautions Regulation 2003

A copy of the Contractor's Health and Safety Policy/ Health and Safety Plan as applicable shall be produced for inspection by the Contract Administrator/ CDM Co-ordinator.

The Contractor shall be responsible for ascertaining whether execution of any order for work complies with their requirements under the CDM Regulations, and must notify the Contract Administrator/ CDM Co-ordinator of same and obtain approval prior to commencement of site activities.

The Contractor will be responsible where applicable, for providing all information required by the Contract Administrator/ CDM Co-ordinator to update the Health and Safety File.

In the event of the Contractor ascertaining that execution of an order will or may involve interference with any hazardous substance or installation then the Contractor shall forthwith notify the same to the Contract Administrator and in so doing shall notify him in writing of any precautions proposed to be taken in consequence of the hazard which may affect the use of the premises or the comfort or freedom of movement of any person likely to be in or near the premises during execution of the order.

The Contractor shall likewise notify in writing the occupant of the premises, or the person in charge of the occupants or users of the premises on which works are in progress or about to be carried out, all restrictions guidance or other precautions which are desirable or necessary for the safety of all persons occupying or using the premises in consequence of the works. The Contractor shall provide all barriers and warning notices required for that purpose and shall make effective arrangements for the occupant or person in charge to consult and communicate with the Contractor, throughout the duration of the works, on the effects and nature of such precautions.

ASBESTOS SAMPLES: in the case of asbestos, the Contractor shall comply with the Control of Asbestos at Work Regulations 2002 and arrange for any necessary sampling and analysis before undertaking any work effecting the suspect material, and shall give the necessary notice to the Health and Safety Executive.

WORKING WITH ASBESTOS: when carrying out work on asbestos based materials, the Contractor shall comply with the Control of Asbestos at Work Regulations 2002 and appoint a licensed specialist sub-contractor. This requirement will be strictly enforced.

Any work with asbestos cement [eg. cleaning, painting, repair or removal]; any work with materials of bitumen, plastic, resins or rubber which contain asbestos, the thermal and acoustic properties of which are incidental to its main purpose and minor work with asbestos insulation, asbestos coating and asbestos insulating board which because of its limited extent and duration does not require a licence (eg. drilling holes, repairing minor damage, painting, removal of angle panel etc.) shall be carried out in accordance with the Approved Code of Practice and Guidance entitled Work with Asbestos which does not normally require a licence (Fourth edition) published by the Health and Safety Executive. The rates for such work shall be agreed with the Contract Administrator.

EMPLOYER'S SAFETY POLICIES: without prejudice to the Contractor's general obligations to ensure compliance with all statutory requirements relating to health and safety, the Contractor shall in particular observe and comply with:

(a) any specific condition, warning or direction given by the Contract Administrator on any matter relating to health and safety;

(b) the relevant provisions of any Employer's Safety Policy applicable to operations of the type in question when undertaken by Employer's employees, being a Safety Policy of which a copy has been given to the Contractor at or before the start of the work:

and

(c) any method statement agreed with the Contractor before the work begins identifying the safety precautions to be taken.

FIRE PRECAUTIONS: the contractor shall take all necessary precautions to prevent personal injury, death and damage to the works or other property from fire. Comply with Joint Code of Practice 'Fire Precaution on Construction Sites' published by the Construction Confederation and the Fire Protection Agency. Smoking will not be permitted on the works.

CANCELLATION ON DEFAULT: in the event of default by the Contractor in the proper observance of any necessary health and safety requirements, cancellation of the written order by the Contract Administrator shall not result in the Employer being obliged to reimburse either any costs incurred by the Contractor or the value of any abortive work except to such extent (if any) as those costs or that abortive work were incurred or performed without contravention of the health and safety requirement in question.

MAINTENANCE OF PUBLIC ROADS: the Contractor shall make good all damage to public roads, kerbs and footpaths, lawns etc, occasioned by exceptional traffic, delivery of materials and building operations generally to the reasonable satisfaction of the Contract Administrator and the local authority.

EXISTING MAINS AND SERVICES: the Contractor shall maintain during the progress of the works, the existing drainage system, water, gas, sewers, electric and other services and is to make arrangements for their continuance and take all necessary steps to protect and prevent damage to them. Should any mains, services ducts or lines be found to be in the way of new works, or require any attention, the Contractor is to seek instructions from the Contract Administrator.

Where it is necessary to interrupt any mains or services for the purpose of making either temporary or permanent connections or disconnections, prior written permission shall be obtained from the Contract Administrator and where appropriate from the local authority or public undertaking and the duration of any interruption kept to a minimum.

PROTECTION, DRYING AND CLEANING: the Contractor shall undertake the following, the cost of which shall be included in the tendered percentage adjustment to the prices in the schedule of rates:-

1. protect all work and materials on site, including that of Sub-Contractors, during frosty or inclement weather.

2. protect all parts of existing buildings which are to remain using polythene/dust sheets and make good any damage caused.

3. prevent damage to existing furniture, fittings and equipment left in the property. Cover and protect as necessary.

4. protect the adjoining properties by screens, hoardings or any other means to prevent damage or nuisance caused by the Works.

5. dry out the Works as necessary to facilitate the progress and satisfactory completion of the Works. The permanent heating installation may be used for drying out the works and controlling temperature and humidity levels, but:

    a) the Employer does not undertake it will be available
    b) the Contractor must take responsibility for operation maintenance and remedial work, and arrange supervision by and indemnification of the appropriate subcontractors and pay costs arising.

6. protect and preserve all trees and shrubs except those to be removed.

7. treat or replace any trees or shrubs damaged or removed without approval.

8. clean the Works thoroughly removing all splashes, deposits, rubbish and surplus materials.

DAMAGE: the Contractor shall exercise great care at all times to prevent damage to the building structure, fittings, furniture, equipment, finishes or the like and shall make good any damage caused by him at his own expense. In this connection all carpets, desks and other furniture or equipment including telephones in the vicinity of the work shall be covered by the Contractor with protective dust sheets or the like prior to any work commencing. Where it is necessary to use any naked flame or welding equipment in executing the work and where combustible materials are in use, adequate protection shall be given to other adjacent materials and personnel. Suitable fire extinguishers shall be provided and readily available at the position where such work is proceeding. The Contractor shall maintain the designated escape routes and exit doors within any building clear of all materials and plant at all times. The Contractor shall consult the Premises Manager at any of the locations in respect of precautions which should be taken for the safety of other occupants prior to the commencement and whilst work is in progress.

PERMIT TO WORK: before commencing any portion of the work the Contractor must establish the need for and if necessary obtain a Permit to Work.

The Contractor's tendered percentage adjustment to the Schedules of Rates shall be deemed to include allowances for time spent in obtaining permits for each portion of the work and claims made by the Contractor shall not be entertained by the Employer in respect of time lost in connection with the issue of permits. However the nett cost of fees and charges by local authorities is reimbursable to the Contractor by the Employer.

Permits to Work will be required for, but not limited to:-

a) Excavation Work
b) Hot Work/ Welding
c) Confined space entry
d) Cutting through or disconnecting existing services
e) Access to roofs

Specific limitations on method/sequence/timing

SITE VISITS: before tendering the Contractor shall visit the site(s) and ascertain:-

1. Local conditions.

2. Means of access to the site(s).

3. The confines of the site(s).

4. Restrictions in respect of loading and unloading vehicles.

5. Factors affecting the order of execution of the work and the time required for the execution of the works.

6. The supply of and general conditions affecting labour, materials and plant required for the execution of the work.

POSSESSION OF THE SITE: no restriction. *

OR

POSSESSION OF THE SITE: possession of the site by the Contractor will be restricted as follows:-

WORKING SPACE : working space is limited to .... *

WORKING SPACE : take reasonable precautions to prevent workmen, including those employed by Sub-Contractors from trespassing on adjoining owner's property and any part of the premises which are not affected by the Works.

* Contract Administrator to complete/delete as applicable

WORKING HOURS: working hours are limited to the normal working hours defined in National Working Rules for the Building Industry as appropriate to the area in which the Works are located. Overtime shall not be worked by operatives on the site without the prior express permission in writing from the Contract Administrator.

OCCUPIED PREMISES: where work is done in occupied premises the Contractor shall take all reasonable care to avoid damaging the property or contents and shall make good all damage which arises from his work.

PROGRAMME OF WORKS: the Contractor shall prepare and submit for the approval of the Contract Administrator, a programme covering all aspects of the works.

SEQUENCE OF WORK OR OTHER RESTRICTION: .... [Note: any restriction on the work (eg. sequence or timing) must be given]. *

USE OF SITE: the site is not to be used for any purpose other than the execution of the contract.

REINSTATE SITE: confine to as small an area as practicable any operations which may affect the surface of the site and reinstate the site after the works are completed.

APPROVAL TO SITING: notify the Contract Administrator of the proposed siting of materials, temporary buildings, rubbish deposits and the like.

TIPPING: no allowance for tipping charges and Landfill tax charges in connection with materials obtained from site including those arising from demolition or alteration works has been included in the schedule of rates and costs should be included in the Contractors tendered percentage 'A'.

OVERTIME: where overtime is ordered in writing by the Contract Administrator, the Contractor will be paid the net additional cost, subject to the addition of 10% for overhead costs provided the contractor's returns are certified each week in relation thereto by the Contract Administrator. ✿

OVERTIME, NIGHT WORK AND INCENTIVES: all costs of overtime or night work at the discretion of the Contractor must be borne by the Contractor and no claims for additional payment in this respect will be allowed.

DAYWORKS: no work will be allowed as daywork unless previously authorised by the Contract Administrator and confirmed in writing. All vouchers specifying the time daily spent upon the work (and, if required by the Contract Administrator, the workmen's names) and the materials used properly priced and extended, shall be signed by the Contract Administrator.

Where daywork is authorised, the Contract Administrator shall be notified of the commencement and completion of the work, and the items of plant and workpeople concerned are to be solely engaged thereon and not employed upon any other work during progress of the daywork.

BUILDING OPERATIONS IN WINTER: the Contractor must be conversant with the measures and operations described in the booklet 'Winter Building' published on behalf of the DOE and obtainable from HMSO for ensuring continuity of work and productivity during inclement weather. The operations and measures described in the booklet shall be taken wherever practicable and having regard to nature, scope and programme of the Works.

✿ This percentage will only apply if no other percentage is inserted in appendix item 12 of the JCT MTC Conditions or is otherwise included in the Contract.

* Contract Administrator to complete/delete as applicable

Facilities/Temporary work/Services

NOTICES AND FEES TO LOCAL AUTHORITIES AND PUBLIC UNDERTAKING: such fees, charges, rates and taxes paid by the Contractor shall be reimbursed nett to him by the Employer. (See also WORKS BY PUBLIC BODIES and PERMIT TO WORK)

WORKS BY PUBLIC BODIES: the ………………………………………….. (name of * local authority or public undertaking) will carry out the following work which will be covered by provisional sums as follows:

Description of Work	Provisional Sum
…………………………………….	……………………………………
…………………………………….	…………………………………… *

The Contractor is to allow for general attendance and overheads.

The …………………………………………. (name of local authority or public undertaking) will be carrying out, in accordance with their statutory obligations, the following work which is not part of the Contract Works: *

Description of Work ………………………………………………… *

[NB: this description should include the scope and timing of the work and its effect on the Contractor's operations]

OFFICES FOR PERSON-IN-CHARGE AND FACILITIES FOR EMPLOYER: the Contractor is to provide, erect and maintain suitably equipped offices for the Person-in-Charge and other necessary staff including Clerk of Works and provide heating and lighting and attendance throughout the duration of the Contract and remove and clear away upon completion and make good all work disturbed.

SITING OF TEMPORARY BUILDINGS ETC: all offices, messrooms, storage sheds, sanitary accommodation and temporary buildings shall be sited to the approval of the Contract Administrator. All areas so used must be made good on completion.

SANITARY ACCOMMODATION: the Contractor is to provide adequate suitable and proper sanitary accommodation which must be water-borne to the satisfaction of the Authority's Health Department and washing facilities for workmen to the standard required by the current Working Rule Agreement and to the satisfaction of the Health Department of the local authority and keep same in a clean and sanitary condition and remove and make good upon completion.

WORKING PLATFORMS: are to be provided to enable the work to be safely and effectively carried out. In all cases the Contractor shall comply with the requirements of the Construction (Health and Safety and welfare) Regulations 1996 together with other relevant regulations and requirements, consulting the Contract Administrator where differing provisions of scaffolding are possible. Agreement by the Contract Administrator to a particular method of scaffolding shall not relieve the Contractor of his responsibility for fully complying with Health and Safety at Work provisions. The Contractor's percentage addition to the prices in the Schedule of Rates is to include for:-

1. temporary staging to provide a working platform up to a height of 1.5 metres.

2. ladders for access to buildings of not more that two storeys in height.

- Contract Administrator to complete/delete as applicable

SCAFFOLDING: scaffolding or towers or mobile towers to provide working platforms greater than 1.5 metres in height will be dealt with as follows:-

1. by the use of the rates for scaffolding contained in Part 2 (Schedule of Descriptions and Prices).

2. by agreement between the Contract Administrator and Contractor.

3. by daywork.

4. by quotations from not less than three specialist firms tendering in competition.

SCAFFOLDING must be constructed in accordance with the requirements of the Health and Safety at Work Act 1974, the Management of Health and Safety at Work Regulations 1992 etc and subsequent amendments or re-enactments thereto and comply with:-

1. BS 5973: 1981 "Access and Working Scaffolds and Special Scaffold Structures in Steel".
2. BS 5974: 1982 Code of Practice for Temporary Installed Scaffold and Access Equipment.

SHORING, SCREENS, FENCING AND HOARDINGS: will be dealt with as follows:-

1. by the use of the rates for such non-mechanical plant contained in Part 2 (Schedule of Descriptions and Prices).

2. by agreement between Contract Administrator and Contractor.

3. by daywork.

4. by quotations from not less than three specialist firms tendering in competition.

NOTICE BOARD: upon written application the Contractor may display and maintain in an approved position a notice board stating his name and that of authorised Sub-Contractors.

PROVISION OF SKIPS: application must be made to the appropriate local authority department for the siting of any skips required for the collection and removal of contractors' waste and rubbish. No allowance for charges in connection with the use of skips has been included in the schedule of rates and costs should be included in the Contractor's tendered percentage. The cost of removal of rubbish on site not arising from the contract works such as fly tipping or spoil from occupants or other contractors is reimbursable.

The Contractor is to ensure that his tender provides for removing rubbish from the site both as it accumulates from time to time and at completion of the works.

Ensure that non-hazardous material is disposed of at a tip approved by a Waste Regulation Authority. Remove all surplus hazardous materials and their containers regularly for disposal off site in a safe and competent manner as approved by a waste regulation authority and in accordance with relevant regulations. Retain waste transfer documentation.

TEMPORARY TELEPHONE: the Contractor is to allow for providing temporary telephone facilities to the site and defray all charges in connection therewith, including the costs of all calls made by his own employees and those of any Sub-Contractors. No provision need be made for telephones for the Employer's representatives.

WATER FOR THE WORKS: the Contractor shall provide all water required for use in the works, by him or by his Sub-Contractors, together with any temporary plumbing, standpipes, storage tanks and the like, and remove on completion. He shall pay all fees and charges in connection therewith and make good all work disturbed.

LIGHTING AND POWER FOR THE WORKS: the Contractor shall provide all artificial lighting and power (electricity and/or gas) for the works, including that required by Sub-Contractors, together with any temporary wiring, switchboards, distribution boards, poles, brackets, etc. and remove same on completion, and pay all fees and charges in connection therewith and make good all work disturbed.

*Note: Where miscellaneous or improvement works to services installations is carried out under a subcontract the services contractor should identify the extent and nature of temporary lighting and power which will be provided by the general contractor at no charge.

METER READINGS: where charges for service supplies need to be apportioned ensure that meter readings are taken by relevant authority at possession and/ or completion as appropriate. Ensure that copies of readings are supplied to interested parties.

Operation/Maintenance of the finished works

Where appropriate the Contractor will provide the Contract Administrator with a free copy of the manufacturers' maintenance/operation manuals for installed equipment.

Submit a copy of each test certificate to the Contract Administrator as soon as practical and keep copies of all certificates.

CONTRACTOR'S GENERAL COST ITEMS

The contractor is to allow in his tendered percentage for the cost of the following:

MANAGEMENT AND STAFF including all overheads, offices, equipment, insurances, travel and expenses, supervision, programming, quantity surveying support staff and the like.

SITE ACCOMMODATION including erection, dismantling, hire charges, maintenance, services, charges, insurances etc. for offices, stores, canteens, compounds, sanitary facilities and the like.

SERVICES AND FACILITIES where not provided by the Employer at no charge to the Contractor, including power, lighting, fuels, water, telephone, security and the like.

MECHANICAL PLANT where not shown as included in Part 2 Schedule of Descriptions and Prices or provided by the Employer at no charge to the Contractor, including cranes, hoists, transport and other mechanical plant.

TEMPORARY WORKS including temporary access roads, hardstandings, hoardings, fans, fences and the like. (The Contract Administrator should define the exact requirements at tender stage)

TRAVELLING AND TRANSPORT OF LABOUR, PLANT AND MATERIALS within the boundaries specified as the 'Contract Area' to the properties listed within the Appendix item 1a including all expenses and vehicle costs.

CLIENT TRAINING: explain and demonstrate to the Employer's premises staff the purpose, function and operation of the installations including all items and procedures listed in the Building Manual.

SPECIFICATION REQUIREMENTS: any adjustments required to the schedule to comply with the specification, drawings or site requirements.

RECORDS AND DRAWINGS of the work as required by the Contract Administrator

GENERAL BONDING AND EARTHING as required by the IEE Regulations for any electrical work

*Contract Administrator to delete/ complete as applicable

IDENTIFICATION AND LABELLING of services specified

BUILDERS WORK INFORMATION, including marking up of any holes or chases, and provision of any sleeving requirements for pipes, conduits, ducts and the like

PAINTING AND PRIMING of services provided

WASTE as generated by the installations including cutting of materials to suit, loss, damage and the like

WORKING in complete conjunction with any ceiling installations

CLEANING of the services works and making good as required

INSPECTION TESTING AND COMMISSIONING of the works. Where testing, balancing and commissioning exceeds the scope of works, the Contract Administrator is to notify the Contractor of his requirements, and an hourly charge or lump sum agreed with the Contractor, if required.

WORK/MATERIALS BY THE EMPLOYER

Plant or materials may be provided by the Employer under the provision of contract clauses 1.6 to 1.12 inclusive.

NOMINATED SUB CONTRACTORS AND SUPPLIERS

Although the JCT Measured Term Contract does not provide for them the Employer may wish to nominate specialist sub-contractors and/or suppliers to carry out works and or supply goods which are not included in the schedule of descriptions and prices. The Contractor will be entitled to 2½% cash discount on the nett cost of the specialist work and 5% on the nett cost of specialist goods after deduction of trade and other discounts, and will also be allowed a percentage addition for profit, attendance and all overheads on Sub-Contractor's accounts after the adjustment to provide the cash discounts described above.**

Attendance includes general attendance and unloading, storing and placing goods in position for fixing, returning crates and packing etc.

(Prices are included for fixing only items which may have been ordered from a nominated supplier or supplied by the Employer). Separate provision shall be made for special attendance in accordance with the Standard Method of Measurement.

WORK BY STATUTORY AUTHORITIES

Where the Contract Administrator orders the Contractor to instruct Statutory Authorities to carry out works under the contract, the Contractor is entitled to recover the full cost of any fees and charges payable in consequence thereof and will be allowed 10% on the net charge to be added for profit, attendances and all overheads.

** This percentage does not apply to Sub Contractors work ordered under the daywork provision in Appendix 11 of JCT MTC conditions unless percentage adjustment for this item has not been included.

Appendix A

List of supplementary clauses which may be applicable on a particular contract.

1. Abortive Calls

2. Administration: Call out procedures
   Reports
   Evaluation
   Complaints procedure, penalties, damages, right of appeal

3. Certificate of Non-collusion

4. Contracting Associations (NICEIC, CORGI and HVCA)

5. Provision of a Contract Bond

6. Communications

7. Data Protection Act

8. Emergency services

9. Employer's obligations and restrictions

10. Good practice and examples

11. Identification

12. Inspections

13. Tenderers proposed method of working

14. Materials, goods and workmanship

15. Notice to occupiers

16. Obligations for statutory tax deduction scheme

17. Annual performance review

18. Potential hazards

19. Protection of furniture

20. Protection of gardens

21. Rehabilitation of Offenders' Act 1974

22. Definition of repairs :   Day to day responsive repairs
    Package maintenance works
    Void property
    Out of hours emergency work

23. Response times

24. Security arrangements

25. Safety of children

Appendix A (Continued)

26. Spot audits

27. Smoking policy

28. Tests

29. Work equipment

30. Use of chlorofluorocarbons

Appendix B

1.0 Transfer of Undertakings and Protection of Employment Regulations

1.1 The present contract is being performed by an outside contractor. The Employer has no view as to whether or not the European Acquired Rights Directive No 77/187 and/or the Transfer of Undertakings (Protection of Employment) Regulations 1981 ("TUPE") applies to this Contract. It is up to each tenderer to form his own view on this. The Employer proposes to ask the present contractor whether he is of the opinion that TUPE might apply to this Contract and if so to provide a list of posts and details relating to it which he anticipates might transfer should TUPE apply. The Employer does not accept any responsibility for whether this information is made available, or if it is correct or not. The Tenderer may be required to complete a confidentiality agreement in respect of this information before it is made available to him.

1.2 The Tenderer must indicate whether his tender is based on TUPE applying or not applying. If the Tenderer has indicated that his tender is based on TUPE applying, he will be taken to have accepted that he will accept a transfer of any staff that should transfer to his contract. If the Tenderer's tender is successful, he must take any issues that may arise about transferring staff (including any questions as to who might transfer to his contract) directly with the present contractor. The Employer will not be willing to become involved in this.

1.3 If TUPE applies the Tenderer should take into account the following requirements in respect of transferring staff,

1. The need to consult with recognised trade unions.
2. The need to maintain existing rates of pay and conditions of employment.
3. The need to provide pension arrangements broadly comparable to those provided at present to transferring employees. (Optional requirement)
4. That liability will transfer to the successful contractor for any claims by transferring employees for redundancy, unfair dismissal or arising out of their previous employment even before transfer.

It will be the Tenderer's responsibility, not the Employer's, to do this.

1.4 The Tenderer is expected in pricing his tender to make his own allowances for and accept the risk of fluctuations in his staffing availability or requirements. The Employer will not accept any tender in which the Tenderer's pricing varies according either to the number, identity or pension status of the staff he requires to perform the Contract or to any changes in the Tenderer's wage rates except so far as they may be reflected directly or indirectly in any method provided in the Contract Conditions for an annual review of the Tenderer's prices.

1.5 If the Tenderer has indicated that his tender is based on TUPE not applying, he must submit with his tender a written statement explaining why he believes TUPE will not apply if his tender is successful. Although this is primarily a matter between the outgoing and incoming Contractor, the Employer is concerned that the transition between contracts should be as seamless as possible. In the interests of continuity if the Employer does not agree with the Tenderer's contention that TUPE does not apply it reserves the right either to reject the tender or to evaluate it according to the Employers own assessment of the financial and other implications based on the Employers' own views as to the applicability to TUPE.

1.6 If the Employer accepts a bid on the basis that TUPE does not apply, it will require the successful tenderer in writing.

1. To accept the risk that TUPE and/or the Directive might apply.
2. To accept that if either is held to apply it will indemnify the Council against any cost that may fall on the Council in respect of any claims under TUPE or the Directive.
3. To agree that if TUPE or the Directive are held to apply, the tenderer will not seek to rescind, repudite, terminate or amend the Contract.

Appendix B (Continued)

2.0   TUPE and the expiry of this contract

2.1   The Employer cannot and does not propose to commit itself as to,

1. What will be its Service requirements after this contract has expired.
2. What arrangements it may propose to make to procure the Service, or
3. What the legislative regime will be at that time either as to procurement of services or transfer of staff.

2.2   It therefore will not enter into any commitment as to what might happen to the successful tenderer's staff at the expiry of the Contract.

SECTION G

MATERIALS AND WORKMANSHIP PREAMBLES

GENERALLY

Unless otherwise stated or contradicted Materials and Workmanship Preambles are to apply reciprocally between Work Groups.

Unless otherwise stated or contradicted the rates contained in Part 2 (schedule of Descriptions and Prices) are to apply reciprocally between Work Groups/ Sections.

The preambles contained in this section are for guidance only, to indicate the basis on which level of pricing has been made. Proprietary brand names have been specified in certain instances. Equivalents (other equal or approved) may be used if approved, in writing, by the Contract Administrator.

For the purpose of these preambles the words "Contractor" or "Sub-Contractor" both mean the person(s) carrying out the works to services installations as a Contractor or Sub-Contractor.

MATERIALS

Where and to the extent that materials are not fully specified they are to be suitable for the purposes of the Works stated in or reasonably to be inferred from the Contract, in accordance with good practice and complying with current British Standards and the recommendations contained within the **current** edition of the CIBSE guide.

Proprietary materials are to be handled and stored strictly in accordance with manufacturer's instructions and recommendations. Such materials are to be obtained direct from the manufacturer's or through their accredited distributors.

Where appropriate, items in the Preambles in other Sections shall apply equally to this Section.

Any preambles included within a specification issued with this document are to apply.

WORKMANSHIP

Where and to the extent that workmanship is not fully specified it is to be suitable for the purposes of the Works stated in or reasonably to be inferred from the Contract, in accordance with good practice and complying with current British Standards and the recommendations contained within the current edition of the CIBSE guide and current IEE regulations. Workmanship is to be of a high standard throughout.

Work liable to damage by frost is not to be carried out at temperatures less than 5 degrees Celsius unless precautions are taken against low temperatures. Submit details of such precautions to the Contract Administrator.

Ensure that site staff responsible for supervision and control of the works are experienced in this type of work.

Items are to include for adequate temporary protection of building structures, decorations, furnishings, equipment and contents from damage by water, dust, debris and the like. Services connections for appliances which are to be removed or relocated are to be cut back and stopped off out of sight wherever possible and left in a neat arrangement.

Naked Flame: where it is necessary to use any naked flame or welding equipment in executing the Work and where combustible materials are in use adequate protection shall be given to other adjacent materials and personnel. Suitable fire extinguishers shall be provided and made readily available at the position where such work is proceeding. Designated escape routes and exit doors, etc shall be maintained and kept clear of all plant and materials at all times.

All reasonable fire precautions shall be taken in respect of stores workshops and other areas/ installations.

Stability of the works: no cutting through floors or walls or under foundations will be permitted other than that required by the drawings or schedules without the sanction of the Contract Administrator. Do not permit anything to be done which may injure the stability of the works or the existing services.

Draining and filling existing systems: where connections are made to an existing installation allowance shall be made for emptying, refilling and venting the existing and new installations during normal working hours.

Allowance shall be made for disconnecting and removing all redundant materials and equipment from site. Such materials shall become the possession of the Contractor and appropriate credit shall be allowed.

Unfinished works shall be left in a safe condition and suitably protected to prevent unauthorised access and interference.

Heavy equipment such as radiators, etc shall be stored in such a manner as to prevent falling or slipping and shall be protected to prevent unauthorised access and interference.

Statutory Authorities: where scope of works affects incoming services, arrange with appropriate Authority to carry out works as necessary to enable the Contract to be carried out.

Burning on site of materials arising from the work will not be permitted without prior approval.

Pipe sizes unless otherwise stated, tubes and their fittings are classified by their internal diameters.

Protection from any variety of damage whatsoever is to be included.

Duct covers, trenches, etc shall be replaced on leaving the site or the immediate vicinity or shall be adequately protected to prevent accidents.

In occupied premises, all works shall be arranged to minimise inconvenience to the normal running of the premises. Any interruptions to the existing services shall take place only with the prior approval of the person-in-charge.

Where thermal insulation to pipework, boilers, calorifiers, etc is suspected of containing asbestos, no works shall be carried out. The Contract Administrator shall be advised of the suspected asbestos and will issue further instructions as considered appropriate.

Workmanship preambles included in a specification issued with this document are to be included.

Delivery periods: immediately upon acceptance the Contractor shall verify the delivery periods of all materials required to complete the works, and must notify the Contract Administrator of any material delivery period which may have an adverse effect on progress.

All investigating work over and above that normally required for installation testing, commissioning, British Standards and Codes of Practice is to be agreed with the Contract Administrator as such prior to execution, and charges on a daywork basis or by a method previously agreed.

FIXINGS TO BUILDING FABRIC: the following shall apply

a) Preparation: mark-out, set out and firmly fix all equipment, components and necessary brackets and supports.

b) Manufacturer's drawings: use manufacturer's drawings and templates for purposes of marking and setting out.

c) Size of fixing: use largest size of bolt, screw or other fixing permitted by diameter of hole in item to be fixed.

d) Greasing of fixings: where indicated, ensure all bolts, screws or other fixings used are greased or suitably lubricated in accordance with manufacturer's instructions.

e) Standards: comply with BS 3974 Part 1 for fixings. Ensure that fixings such as expanding anchors are tested for tensile loading with BS 5080.

f) Plugs: use plugs of suitable size and length for fixings. Use plastic, fibrous or soft metal non-deteriorating plugs to suit application. Do not use wood plugs.

g) Screws: use screws to BS 1210. Generally use sheradized steel wood screws for fixing to concrete, brickwork or blockwork. Grease screws where indicated. In damp or exposed situations use greased brass wood screws.

h) Shot fired fixings: obtain approval prior to use of shot fired type fixings.

i) Self adhesive fixings: use self adhesive type fixings where indicated.

j) Drilling: drill holes vertical to work surfaces. Use drills of requisite size and depth, and appropriate to fabric. Flame-cut holes in metal work are not permitted.

k) Fixing to reinforced concrete: take precautions to avoid fixing through reinforcement.

l) Fixing to brickwork: do not fix to unsound material or mortar between brickwork courses.

ELECTRICAL SERVICES

MATERIALS

LV SWITCHGEAR AND DISTRIBUTION BOARDS:

GENERAL STANDARD: Comply with BS 5486.

ENCLOSURES	:	ensure enclosure provides minimum degree of protection indicated
Door fastening	:	supply doors with fastenings and provision for locking in closed position
Covers	:	use covers which require special tools for removal
Fixings	:	provide enclosure with fixing holes as specified. Where enclosure is located outside provide fixing lugs external to enclosure
Earthing terminals	:	fit enclosures with earthing terminals suitable for internal and external connection, so that exposed conductive parts of factory built assembly can be connected to protective conductor
Marking	:	screw labels to outside of switchboards

ENCLOSURES FINISH: apply high standard finish to enclosure and supporting metalwork. Degrease metal and remove rust prior to applying finish. Finish and colour as specified.

TERMINALS FOR EXTERNAL CONDUCTORS: make terminals for neutral on three-phase and neutral circuits same size as phase terminals, unless otherwise indicated.

TERMINAL BLOCK FOR AUXILLARY WIRING: provide railmounted moulded terminal blocks with fully shrouded connectors. Provide connectors to clamp conductors between metal surfaces. Provide each terminal with marking tag fitted into moulded tag slots.

SWITCHBOARDS: supply switchboards comprising assembly of switchgear, controlgear and components as indicated: BS 5419, 5420 and 5846.

External	1	design as specified
	2	degree of protection as specified
	3	steel enclosure or as specified
	4	degree of accessibility as specified
	5	earthing as specified

Neatly arrange and securely fix wiring for switchboard components, instruments, meters, controls and interlocking. Where appropriate protect by cartridge fuses complying with BS 88.

AIR BREAK SWITCHES AND FUSE SWITCHES: comply with BS 5419. Supply air break switches with uninterrupted rated duty. Endurance as specified. Fit each switch with facility to padlock in OFF position.

PADLOCKS: provide each switchboard with sets of padlocks as indicated.

FUSES: supply cartridge fuse links including fuse carrier, bases and associated components that comply with BS 88, fusing factor Class Q1, unless otherwise indicated.

MCCB's: comply with BS 4752.

MCB's: comply with BS EN 60898.

DISTRIBUTION BOARDS: comply with BS 5486 Parts 11 or 13, as appropriate. Make internal separation Form 1 unless otherwise indicated. Make fuseboards fully shrouded. Install busbars in same position relative to their fuse carriers or miniature circuit – breakers (MCBs) for each pole. In TPN distribution boards supply neutral busbars with one outgoing terminal for each outgoing circuit.

Provide a multi-terminal earthing bar for circuit protective conductors for both insulated and metal cased boards, with one terminal for each outgoing circuit. Identification as specified.

CABLE TERMINATIONS: ensure that switchgear and distribution boards are provided with facilities to terminate size, number and type of cable indicated. Where necessary use fabricated steel extension boxes for glanding plates and multiple cables. Provide non-ferrous metal glanding plates for single core cable terminations.

LIGHTING:

Standards	:	comply with BS 4533 and BS 5225
Emergency Lighting	:	comply with ICEL:1001 or as specified
Luminaires		
Exit signs	:	comply with BS 5499-3
Hazardous area	:	comply with BS 889, BS EN 50014, BS 5345 and BS 5501 as applicable
Signs & High Voltage Lighting		
Installations	:	comply with BS 559
Classification	:	comply with BS 4533 Part 101, as indicated
Flammability of translucent covers	:	submit a certificate of tests carried out to BS 2782, where a specific degree of flammability in indicated
Safety	:	fit luminaire with cover glass to protect against ultra-violet emission and risk from explosion of lamps, where indicated
Safety support	:	provide secondary support for translucent covers, diffusers and gear for components trays so they are prevented from falling when their primary fixing is released
Photometric performance	:	ensure luminaries of similar type have some photometric performance as published data within the tolerances defined by BS 5225

LAMPHOLDERS

Standards	:	comply with BS EN 61184 and BS 5101 as applicable
Interchange-ability	:	ensure lampholders in luminaries of similar type and rating are identical
Tungsten fitting	:	use following lampholders for tungsten filament lamps unless indicated otherwise:-

LAMP	LAMPHOLDER
Up to 150W	Bayonet B22d
200W	Edison screw E27
300W and above	Edison screw E40

Mounting	:	securely mount lampholder in luminaire when it is sole support for lamp
Cord grip	:	provide integral cord grip type when lampholders are suspended by cord
Conduit mounted	:	when mounted directly to conduit system use backplate lampholder suitable for conduit box
Shade rings	:	provide a shade carrier ring for separately mounted lampholders for GLS tungsten filament lamps
Earthing	:	ensure metal lampholders incorporate an earthing terminal
Polarity of Edison screw lampholders	:	ensure phase conductor is connected to centre contact

CONTROL GEAR AND COMPONENTS:

Fluorescent Lamp

Ballasts	:	comply with BS 2818. Use low distortion type

Discharge Lamp

Ballasts	:	comply with BS 4782
Capacitors	:	comply with BS EN 61048 and BS EN 61049 where applicable
Supply terminals	:	use screw terminals for supply cables and circuit protective conductors, sized to terminate up to three 2.5mm^2 conductors. Provide separate terminal blocks for each incoming circuit, with marking to identify each circuit
Fuse	:	include a fuse holder and BS 1362 fuse in each incoming circuit phase connection
Compatibility	:	ensure control gear and components are suitable for lamp type, wattage and starting characteristics
Interference	:	comply with BS 5394

Remote Gear	:	where indicated, locate control gear in separate lockable cabinet of sheet steel with same degree of protection and finish specified for luminaire. Comply with manufacturer's recommendations for maximum cable length between gear and lamp

LAMPS:

Tungsten filament lamps	:	comply with BS 161, BS 5971 and BS 6179 as applicable
Fluorescent lamps	:	comply with BS 1853
High pressure mercury vapour lamps	:	comply with BS 3677
High pressure sodium vapour lamps	:	comply with IEC 662
Low pressure sodium vapour lamps	:	comply with BS 3767
Manufacturer	:	ensure that lamps of each type are from the same manufacturer

SUPPORT SYSTEM:

Conduit	:	use not less than 20mm conduit of same type as main conduit system. Material as specified
Rod	:	use 10mm diameter continuously threaded rods with matching washers and nuts. Material as specified
Chain	:	use cadmium plated steel chain with load carrying capacity of not less than twice weight of complete luminaire, or as specified
Flexible cord	:	confirm temperature rating is suitable for operating temperature of luminaire or lampholder
Wall brackets	:	confirm wall brackets are suitable for supporting luminaire
Ball & socket	:	provide ball and socket complete with cover fixed to circular conduit box, as top support where indicated

COLUMNS AND BOLLARDS:

Materials type	:	as indicated
Finish standards	:	comply with BS 5649 as applicable
Service compartment	:	provide compartment to take cable, terminations, fused cut-out, control gear and other items as indicated. Make door lockable, with degree of protection as indicated
Bracket	:	match column. Type and length as indicated
Wiring	:	ensure cabling between control gear and lamps complies with lamp manufacturer's recommendations for voltage drop. Confirm temperature rating is adequate for operating temperature of luminaire
Earthing	:	include earthing terminal fixed within service compartment
Base	:	provide as indicated

ELECTRICAL ACCESSORIES:

Interior lighting switches : comply with BS 3676

1. switch type, current rating and earthing requirements as specified
2. enclosures, finishes and ancillaries as specified

Ceiling roses : comply with BS 67

Lampholders : comply with BS 5042 T2 rated

Fused connection units : comply with BS 5733

Switched fuse connection units : comply with BS 5419 and BS 5846

1. suitable for conduit or gland entries
2. HBC fuses to BS 88 or BS 1361 or as specified

Switched socket outlets : comply with BS 1363

1. 3 amp rating
2. type and earthing requirements as specified
3. enclosures, finishes and ancillaries as specified

Double pole control switches : comply with BS 3676

CONDUIT AND CABLE TRUNKING:

CONDUIT AND FITTINGS: comply with BS 31, BS EN 60423 and BS EN 50086-1. Use conduit from one manufacturer throughout. Do not use inspection bends or couplers except with 32mm diameter conduit.

RIGID STEEL CONDUIT: use seam welded steel screwed conduit and couplings to BS 4568, Part 1, heavy gauge or as specified.

Finish	:	as specified
		Ensure fittings are same class and finish as associated conduit system
Adaptable boxes	:	use covers of same material as adaptable and conduit boxes
Conduit boxes and covers	:	secure boxes with malleable iron or flat steel covers by steel dome or cheese headed screws for Class 2 finish or brass dome or cheese headed screws for Class 4 finish
	:	use purpose made boxes at right angle changes of direction
	:	limit number of entry holes within loop in boxes to four
	:	use 100mm x 100mm x 50mm minimum size adaptable box unless otherwise indicated
	:	use couplers and externally screwed brass bushes to connect conduit to loop-in circular conduit boxes
	:	use washers with flanged couplers
Plugs	:	as specified

FLEXIBLE AND PLIABLE STEEL CONDUIT:

Type	:	flexible conduit: comply with BS 731 Part 1
Protection	:	as specified
Fittings	:	supply fittings of same class and finish as conduit system
Adaptable boxes, conduit boxes, covers and plugs	:	as for steel conduits
PVC CONDUIT	:	comply with BS 4607
	:	accessories same manufacturer as conduit
	:	accessory boxes comply with BS 4662 or BS 5133

TRUNKING AND FITTINGS: supply partitions and covers of same material as trunking, unless otherwise indicated. Ensure gap between partitions and lids is minimum to maintain segregation of circuits.

Use standard fittings where possible. Where site fabricated fittings are necessary ensure they are comparable in construction and finish with system.
Take measurements on site before producing drawings for manufacture of trunking

Electrical continuity for steel trunking	:	maintain electrical continuity at each joint by a copper link, tinned copper for galvanised trunking, fixed on outside of trunking, secured by screws, nuts and shakeproof washers. Make provision for continuity to be achieved without need to remove paint from ferrous metal where trunking has a painted finish
Access	:	ensure that when open a minimum of 80% of normal trunking width is available for access
Cable supports	:	provide horizontal trunking with removable bridges to retain cables in situ
	:	provide vertical trunking with pin racks to support cables at 1200mm maximum spacing
	:	use insulated pins or insulation sleeved pins on pin racks
Connection	:	use purpose made fittings where trunking connects to switchgear and fuseboards
	:	ensure rigidity of trunking is maintained across connection
Dimensions	:	ensure external dimensions of trunking are maintained across connection between trunking lengths
	:	ensure internal dimensions of trunking are not reduced by more than 4% across connection

SURFACE STEEL TRUNKING:

Material	:	sheet steel trunking comply with BS 4678 Part 1 and BS EN 10143 or as indicated
Trunking type	:	as specified
Gauge of metal	:	1.2mm sheet steel metal for trunking up to 75mm x 75mm or equivalent cross sectional area
	:	1.6mm sheet steel metal for trunking above 75mm x 75mm or equivalent cross sectional area
Degree of protection	:	as specified
Finish	:	as specified
Colour	:	as specified
Fixings	:	provide external fixing lugs where specified protection is IP44 or greater
Jointing	:	make joints in trunking and fittings with connectors secured with screws, nuts and shakeproof washers
Style	:	use trunking manufactured with inward return edge flanges and fitted with flange couplers
Fittings	:	use bends, tees and angles of similar gauge, type and finish as trunking body and supplied by same manufacturer
	:	obtain approval to site fabricated special accessories
Lighting trunking:		provide steel trunking to comply with requirements for lighting trunking
	:	surface with snap-on lid
	:	provide cover strip to prevent ingress of foreign materials, locate cabling in place and insert closure strips between fluorescent luminaries. Use trunking cover strip clipped into place in trunking body. Cover strip material:- galvanised steel or PVC as specified earthing as specified

SUPPORTS AND FIXINGS: provide proprietary suspension systems comprising channel sections with return lips and comparable fixing accessories to BS EN 10162 or BS EN 10210-2 and/ or slotted angles to BS 4345.

Ensure support components for Class 4 conduit are hot dip galvanised after manufacture. Ensure components in direct contact with conduit match profile of conduit. Ensure all studding, U bolts and steel screws, bolts, nuts and washers are either cadmium or zinc electroplated to BS 3382 after manufacture. Do not use metal fixing components likely to cause damage through electrolytic action.

CABLE TRAY:

General	:	cable trays shall be manufactured from sheet steel to BS 1449. The thickness of the sheet shall be 1.2 metric gauge up to 225mm wide, 1.6 metric gauge for sizes from 225mm to 450mm, and 2.0 metric gauge for sizes wider than 450mm.
	:	cable trays shall have solid flanges at right angles to the tray to provide rigidity. All bends and tees shall be manufactured standard accessories where possible. A copper earth bonding strip shall be fitted at each joint to ensure earth continuity.
Finish	:	cable trays shall have a galvanised finish, to BS 723 after manufacture.
General installation	:	Where multiple runs of small cables are required, particularly when mineral insulated cables are in use, they may be run on perforated cable tray.
	:	cable shall be fixed to tray work with saddles of the type and spacing set out in the sections covering the respective cables.
Supports	:	cable tray may either be run horizontally and supported from soffits or cantilevered wall brackets by means of steel rods of not less than 6mm diameter, or supported directly on top of cantilever brackets, or mounted vertically as required. Fixings shall be such that there will be no perceptible deflection on the trays when all cables are in position, and shall not exceed 1500mm centres.
	:	where fixed directly to horizontal or vertical surfaces cable trays shall be spaced off the surface using unichannel sections.
Protection of cables	:	where cables enter or leave cable trays the Contractor cables shall ensure that no sharp edges are carrying the weight of cables and that no cables subject to vibration are in a position where they could be abraded by edges of the cable tray.
HV/LV cables and wiring	:	comply with BS 729
Impregnated paper insulated cables	:	comply with BS 6480
PVC insulated armoured and unarmoured cables	:	comply with BS 6346 for PVC and BS 5467 for XLPE
PVC/ PVC cables	:	comply with BS 6004
Rubber insulated LV cables	:	comply with BS 6004, BS 6007, BS 6500 or BS 7919
Flexible cables	:	comply with BS 6500
Thermoplastic insulated single core	:	comply with BS 6360

Mineral insulated cables	:	comply with BS 6207, PVC sheathed where specified
Cable glands	:	metric dimensions to BS 6121, imperial dimensions to BS 4121

1. seal as specified

2. earthing and bonding as specified

CABLE SUPPORTS: support all cables throughout their length using:-

    conduit, trunking and enclosures
    cable tray, rack or special support systems
    cleat or clip direct to building fabric
    aerial catenary suspension systems

Ensure tray, racking and special support systems are continuous and firmly fixed to the building fabric. Allow space for cables.

Ensure cable spacing complies with IEE Regulations without derating.

Cable cleats   :   material and type as specified

CABLE SUPPORT SYSTEM FINISHES: for all support components, fixings, hangers and accessories use

    galvanised, sheradized finish, or mild steel, painted with red oxide, cable cleats, material and type as specified

FIRE ALARMS: the fire alarm and detection system shall be in accordance with the following standards, rules, recommendations and requirements as appropriate:-

- BS 5839
- BS 3116
- BS 5364
- BS EN 54-1
- BS 6266
- CP 1022
- The IEE Wiring Regulations
- FOC Rules
- Local District Surveyor or Building Control Officer
- London Fire and Civil Defence Authority or Local Fire Brigade

EXTRACT FANS: construct casing entirely rigid and free from drumming under bearing support, as indicated, mild steel to BS 1449 Part 1 or as specified

1. motor speed – single, variable or two

2. drive – direct or belt driven

3. accessories as specified

MOTOR CONTROL AND CONTACTOR SYSTEMS:

1. switches and relays are to comply with BS 4794

2. starters to comply with BS 4941

3. motors to comply with BS 4999

EARTHING AND BONDING: Earthing systems are to comply with IEE Regulations where appropriated British Standard Code of Procedure BS 7430

LV Switchgear and Distribution.

HBC fuses:

Fitting in fuseboard	Purpose-made fuseboard with spare ways
Isolators:	Internal standard, not weatherproof
Switched fuses and fused switches:	Internal standard, not weatherproof
Distribution boards and consumer units:	Fixing to brickwork/ concrete, etc using self-drill anchors, bolts and washers. Internal standard, not weatherproof

Lighting Installation

Luminaires:

Tungsten/ fluorescent/ discharge/ Emergency fittings:	All light fittings based on "Thorn" range
Bulkhead type:	Fixing to brickwork/ concrete using plugs and screws
Recessed downlight/ recessed fitting:	Fixing within lay-in ceiling grid, includes for plug in ceiling rose
Batten fitting:	Fixing to concrete soffit using plugs and screws
Pendant fitting:	Fixing to existing cables
Lighting switches:	Based on "MK" range

Power Installation

Connections to equipment supplied and fixed by others:	Up to and including 16mm^2 cables including 1m kopex and including glands
Switches/ sockets:	Based on "MK" range. New back box

Mechanical Services Wiring and Controls

Connections to equipment supplied and fixed by others:	Up to and including 4mm^2 cables
Starters, isolators, stop buttons:	Internal standard, not weatherproof
Thermostats/ frost-stats/ temperature detectors:	Up to and including 4mm^2 cables
Tubular heaters fix and connect only equipment supplied by others:	Based on "Bush Nelson" range
Tubular heaters, hand driers, water heaters:	Fixing appliance to wall "Satchwell" range

Mechanical Services Wiring and Controls (Cont'd)

Thermostats:
Extract ventilation system: "Xpelair" Fans

Air temperature detectors,
room thermostats, manual control: Wall mounted at 1.5m from FFL

System of Wiring

Conduit: Includes for 1 saddle and 1 coupler per m. Based on "Walsall" range. Fixing to concrete soffit, maximum height 3m

                                                         Flexible conduits; "Adpataflex"

LV Switchgear and Distribution

Distribution boards and consumer units: Internal standard, not weatherproof

                                                         Equipment based on "Crabtree" and "MEM" ranges

Power Installation

Electrical connections: Up to and including 4mm^2 cables

Switches, socket outlets,
fused connection units: Recessed pattern, concealed conduit installation: new back box. Based on "MK" range

Electric Heating Installation

Electrical connections: Up to and including 16mm^2 cables

Water heaters: Based on "Santon" range

Immersion heaters: heat emitters: Based on "Santon" range

Thermostats:

Room-stats: Standard finish. Based on "Satchwell"

Frost stats: Externally mounted

Extract Ventilation Installation

Electrical connections: Up to and including 16mm^2 Cables

Extract fans: Includes for flexible connection. "Xpelair" fans

Motor Control and Contactor systems

Electrical connections:	Up to and including 4mm^2 cable
Starters/ motor control switches:	Internal standard, not weatherproof, for installation in MCC's – no enclosures
Stop buttons:	Oil proof type
Contactors:	As starters
Motors:	Free-standing motor, with belts/ pulley/ guard. Includes re-tensioning belts
	Equipment based on "Crabtree" range and "MEM" ranges

Fire Alarm Systems

Electrical connections:	Up to and including 4mm^2 MICC
Equipment:	Based on "Menvier" range

System of Wiring

PVC 300/ 500 V:	Includes for clips/ cleating and or reistatement of trunking lid. All cables based on BICC
MICC:	Includes for clips

Earthing and Bonding

excludes cable termination

System of Wiring

Conduit:	Includes for 1 saddle and 1 coupler per m. Based on "Walsall" range. Fixing to concrete soffit, maximum height 3m
	Flexible conduits; "Adaptaflex"
Trunking:	Includes for couplers and lid and hanger only. Based on "Salamander" range. Fixing to concrete soffit, maximum height 3m
Adaptable boxes:	Includes attaching of conduits/ trunking etc
Channel:	Includes fixings
Capping:	Includes fixings
Tray:	Includes for couplers. Based only on "Swifts" range fixing to wall
Cables and conductors, cables:	Laying on pre-fixed tray and fixing in position using cable clip/ cleats. Tagging cables

ELECTRICAL SERVICES

WORKMANSHIP

LV SWITCHGEAR AND DISTRIBUTION BOARDS:

Fixings and mounting requirements	:	as specified
Earthing and bonding	:	as specified
Circuit breakers	:	comply with BS 3871/ 4572 and type as specified
Wiring requirements to distribution boards	:	as specified

CABLE TERMINATIONS: terminate paper-insulated cable by means of switchboard manufacturer's standard compound filled cable boxes. Terminate PVC SWA PVC and MICC cables inside enclosure by securing cables to switchboard with glanding plates or glanding brackets; and outside enclosure with glanding plates or fabricated steel extension boxes.

LIGHTING INSTALLATION:

Orientation	:	install luminaries as indicated, and in horizontal plane unless otherwise indicated
Cleanliness	:	ensure luminaires are clean and grease free on handover
Recessed fittings	:	install luminaries flush with finished ceiling level
Semi-recessed fittings	:	install luminaries as manufacturer's detail or as indicated
Wall mounted fittings	:	install luminaries at height indicated
Material of supporting surfaces	:	ensure classification of luminaires is appropriate. Do not mount luminaires on readily flammable surfaces
Potentially explosive atmospheres	:	comply with BS 5345
Signs & high voltage installation	:	comply with BS 559

SUPPORT:

Type : use type of support as indicated. Ensure support is adequate for weight of luminaires

Number : provide not less than following number of supports for each luminaire longer than 600mm:-

Luminaire Width mm	Minimum number of supports
Up to and including 300	2
Over 300	4

Support : where luminaire is supported from conduit provide a conduit box forming an integral part of conduit system at each point of suspenion

: where luminaire is supported from trunking use proprietary clamps or brackets appropriate to the luminaire and trunking

: other methods of support as specified

SUSPENSION

Appearance : suspend luminaires at height indicated. Ensure suspensions hang vertically unless otherwise indicated

Conduit : where conduit enters luminaires use back nuts and washers to secure luminaire body to conduit support. Provide tube with corrosion resistance equal to conduit system.

Rod : use washers, nut and lock-nut at top and bottom or rod. Paint cut ends with calcium plumbate primer or zinc rich paint

Chain : use hook cover for suspension from circular conduit box. For connection to luminaire use manufacturer's own chain hook, but if not available use hook with standard screw threaded body to be secured to luminaire body with nuts and washers. Where indicated use captive hooks

Flexible cord : suspend cord from ceiling rose

Ball and socket : install cable through ball and socket connected to conduit box

Wiring : shall be carried out on the "loop in" basis, and all sub-circuit cables shall terminate in the ceiling spaces as specified

1. method of connections to luminaires as specified

2. earthing and bonding as specified

3. identification as specified

ELECTRICAL ACCESSORIES: fixed equipment, water heaters, immersion heater, storage heaters, cookers etc., shall be installed on separate radial circuits as per IEE regulations.

EARTHING: ensure metal framework of equipment is bonded to main earth point. Ensure that cable CPC's are connected to earth bar. Provide earth bar CPC between earth lug on metal box and accessory casing except where accessory is encased in plastic.

PROTECTION: ensure there is no physical or electrical damage to accessories when they are removed from their packaging and during installation. Provide masking covers for surface mounted accessories to protect surface from paint where accessories are flush mounted install front plate after painting is finished.

FIXING: align accessories horizontally and vertically, as indicated. Where accessories are grouped, mount horizontally in line and parallel to each other and equidistant. Fix cover plates to boxes with brass fixing screws.

CONDUIT AND CABLE TRUNKING:

GENERAL: arrange conduit, trunking and ducting to present neat appearance, parallel with other service runs and lines of building construction, except where in screed or in-situ concrete. Ensure plumb vertical runs.

Install cable in conduit, trunking or equipment enclosures throughout its length for protection. Do not use framework of partitions or similar unless indicated.

Make provision in conduit and trunking at expansion and settlement joints to allow for movement of building structure. Use manufactured expansion couplings for conduit and trunking.

Provide adaptable boxes 300mm either side of expansion or settlement joints for conduit crossing.

Join boxes with either:-

- : flexible steel conduit type as indicated
- : conduits arranged to form a telescopic joint and cover overall with PVC sleeve to provide minimum degree of protection of IP44

Ensure entire system is electrically and mechanically continuous.

LAYOUT: ensure maximum circuit lengths and groupings of cables indicated are not exceeded. Where dimensions are not indicated select trunking sizes to allow for additional cables up to limit consistent with grouping specified for cables to be accommodated.

SPACING: install steel conduit, metal trunking and equipment clear of other services. Measure distance from external surface of any lagging. Notify instances where minimum clearance cannot be achieved and bond items concerned.

Minimum general spacing between conduits, metal trunking and equipment and:-

- a : insulated steam services – 300mm
- b : other services excluding steam – 150mm
- c : above central heating radiators – 1000mm

CONDENSATION PREVENTION: install conduit and trunking systems to ensure internal condensation does not affect operation of associated circuits. Comply with IEE Regulations for provision of drainage points.

Where conduit passes through external wall between two areas of different ambient temperatures or in other locations likely to cause condensation, install conduit box. After wiring, fill box with inert, permanently plastic compound with high insulation value.

SCREWED STEEL CONDUIT: use materials clean and free from defects and, where ferrous materials are involved, free from rust, scale and oil. Obtain consent for use of materials subject to remedial work. Install conduit throughout of sufficient cross sectional area to allow cables to be easily drawn in or out singly or bunched.

Cut conduit clean and square with axis. Remove any metal burrs prior to installation.

STEEL CONDUIT FITTINGS: site form bends in conduit wherever practical. Do not use manufactured bends, elbows or tees except where indicated.

Construct bends and sets cold with a bending machine without altering section of tube. Do not apply heat when forming sets or bends.

Use bending tools complying with British Standards appropriate to conduit material. Avoid marking or damage to conduit components.

Ensure length of thread on conduit matches that in conduit fittings and equipment with no thread exposed after erection except at running couplers.

Use lubricant when cutting threads.

Ensure conduit butts in couplers. Use minimum number of running couplings.

    a.   For running couplings in Class 2 conduit, use coupler and locknut. Paint exposed thread with zinc rich paint

    b.   For running coupling in Class 4 conduit, use three piece conduit unions

DRAW-IN BOXES: provide draw-in boxes in conduit at maximum intervals of 10 metres.

PVC CONDUIT: shall be installed in accordance with the manufacturer's recommendations and the IEE Regulations.

INSTALLATION OF CONDUIT IN SCREED: check there is no mechanical damage to conduit in floor screed prior to screeding. Fix securely before screed is poured. Provide temporary protection to conduits until screeds are laid. Ensure there is no blockage immediately shuttering is removed.

Ensure minimum amount of cross overs occur dependent upon screed depth. Do not install draw boxes in floors.

Do not install conduits:-

    a.   in screeds in areas indicated

    b.   within site blinding

    c.   in main structural slabs unless permission is obtained

CONDUIT BOXES: fit circular conduit boxes with extension rings where terminal blocks are accommodated.

Ensure fixing holes are countersunk to prevent screw heads projecting into boxes and remove burrs before cables are drawn in. Use two minimum screw fixing for standard circular conduit boxes and four screws for large conduit boxes and adaptable boxes up to 150mm x 100mm.

Use back outlet boxes where surface conduits pass through walls, to outside accessories or lighting points.

Secure switch boxes and socket boxes using countersunk steel screws and plug inserts and finally grout in position prior to plastering or screeding.

WIRING: comply with IEE Regulations when wiring installations. Segregate circuits as indicated.

Ensure draw wires left within empty conduits for use of specialist installers. Use draw wires as specified.

For concealed conduit ensure system is installed to enable re-wiring to be carried out from fittings, boxes and switch boxes only. Draw-in boxes will only be permitted in special instances.

FIXING CONDUIT: support conduit in accordance with IEE Regulations, Table 11C. Ensure conduit is not under mechanical stress. Fix conduit boxes independently of conduit. Make allowance for any additional mechanical loading supported by conduit boxes.

Where protection is specified as IP44 or greater ensure fixings of conduit boxes are suitable to maintain degree of protection.

Use following methods of fixing conduits:-

LOCATION	TYPE OF FIXING
Floor screeds	Saddles or crampets
Buried in plaster	Crampets or saddles or render
Above false ceilings	Spacer bar saddles

BUILDERS WORK: ensure conduit is not concealed until work has been inspected and approved. Obtain permission before horizontally chasing walls. Ensure that conduit and fittings buried in concrete or behind plaster are protected against corrosion or electrolytic action prior to rendering.

Ensure conduit concealed in wall chases is covered in full plaster thickness and/ or rendering to minimum depth of 12mm.

Use extension collars of appropriate depth to leave boxes flush with finished wall and ceiling surfaces.

FLEXIBLE AND PLIABLE CONDUIT: use flexible conduit for final connections to motors and other equipment subject to vibration or adjustment and thermostats, motorized valves and similar items mounted in pipelines or ducts.

Use sufficient length between equipment and box at end of conduit run (minimum 450mm) to allow necessary full range of withdrawal, adjustment or movement.

Use clamp type adapters to terminate flexible conduit.

Install flexible conduit to ensure conduit will suspend naturally with no stress due to bends or sets being creased against position of building fabric.

Use PVC covered flexible conduit where installed externally, exposed to weather or in any position where ingress of moisture or condensation may occur.

STEEL TRUNKING: install steel trunking in accordance with IEE Regulations. Use trunking to avoid multiple parallel conduits runs, subject to approval.

Cut trunking clean and square with axis, prepare ends and remove burrs and sharp edges. Ensure inside of trunking is free from anything liable to damage cables either during installation or after covers are fitted. When trunking is held in a vice, ensure surfaces remain undamaged and components are not warped. Avoid tool marking or damage to trunking system components.

Use folding bars when bending trunking fabric to site form junctions and angles. Ensure corners are neat and metal on either side of corners is not distorted. Do not form flanges by cutting or bending trunking material. Form circular holes in trunking body using correctly sized punch sets. Use twist drill for holes 6mm maximum diameter.

Use only factory formed openings for accessories.

Line unprotected apertures in trunking with PVC or nylon edging strip.

Fit ends of runs with removable blanking off covers unless indicated or approval obtained.

Provide fixed section of cover projecting 25mm either side of fabric where trunking passes through wall, floors or ceiling.

Fit cable retaining straps at 500mm intervals except where cover is on top.

PVC TRUNKING: shall be installed in accordance with the manufacturer's recommendations and IEE Regulations.

FIXING TRUNKING: ensure trunking is independently fixed and supported from building fabric. Obtain approval for proposed fixings/ supports. Ensure trunking is supported to obviate excessive deflection both in vertical and horizontal planes.

Support trunking in accordance with IEE Regulations Table 11D and:-

: two fixings minimum per standard length
: as indicated

FIRE BARRIERS: ensure trunking passing through fire barrier floors or walls incorporates barrier of fire resisting material at each point to prevent spread of fire.

ACCESS: arrange trunking to allow access to wiring. Locate covers on top or sides of trunking if practicable. Notify where this cannot be achieved.

PROTECTION AND REPAIR OF STEEL COMPONENTS: paint joints of conduit and minor damages to finish of conduit and trunking immediately after erection or after damage occurs.

Use paint compatible with finish. Remove grease, oil, dirt and rust before applying protective paint.

Notify serious damage and repair or replace as instructed.

CLEANING BEFORE WIRING: clean inside of conduits and trunking with swabs immediately before wiring.

Inspect all components and remove any foreign matter, fit temporary plugs to open ends of conduit and trunking to prevent ingress of water and solid material.

EQUIPMENT CONNECTION: where surface mounted equipment is installed in conjunction with concealed conduit work, terminate concealed conduit at flush mounted conduit or adaptable box. Drill back of equipment, bush for back entry and mount equipment to conceal back box.

Connect to fixed equipment via conduit box located adjacent to termination point, using either solid or flexible conduit as indicated for final connection to equipment terminations.

Use conduit box as cable change point to facilitate changed wiring locally to adjacent equipment.

Connect trunking to equipment by specially fabricated connectors or by couplers and externally screwed brass bushes.

CABLE TRAY: the system shall be electrically and mechanically continuous bends being such that the minimum bending radius of the largest cable to be installed is as specified. Where cable trays are exposed extraneous metal work they shall be bonded in accordance with IEE Regulations (i.e. supplementary bonding). The tray shall be supported by a maximum of 3m apart or as specified by rod or steel angle iron painted before or immediately after erection.

Spacing for cable fixings to be as table 11A of the IEE Regulations.

HV/ LV CABLES AND WIRING:

CABLE MANUFACTURER: use new cables delivered to site with seals intact, manufactured not more than one year prior to delivery, labelled with manufacturer's name, size, description, BS number, classification, length, grade, and date of manufacture.

CABLE CERTIFICATION MARKING: mark all types of cable included in British Approvals Services for Electric Cables (BASEC) in accordance with BASEC regulations or with equivalent CENELEC HAR cable certification marking.

CABLE INSTALLATION – GENERAL: lay cables in one length. Obtain permission from Contract Administrator for all through joints, and where overall length requirement exceeds practical drum size.

Install cables only when ambient temperature is 0°C or greater, using cables stored at or above this temperature for not less than 24 hours, unless a written statement is produced by the cable manufacturer that a lower installation temperature is acceptable. Use special tools where recommended in manufacturer's installation instructions.

Clearances and spacing of cables to be as specified.

Routing and protection to be as specified.

CABLE INSTALLATION IN DUCTS: prove ducts clear and clean out, by drawing a mandrel, size 150mm long, diameter 12mm less than duct bore, followed by a circular wire brush, diameter 12mm more than duct bore, through each duct immediately prior to pulling in cables.

Provide all axles and axle stands, fair leads, rollers, cable stockings and other equipment necessary to avoid cable abrasion damage during installation. Use only lubricants recommended by cable manufacturer to have no deleterious effect on cable to assist drawing process.

Ensure bending radium of cable during drawing is not less than permanent installation. If drawing single core cables forming a trefoil, bind together with PVC tapes at 1m intervals, and draw in as one cable.

Seal between cable and duct to prevent the ingress of water after cable installation.

CABLE INSTALLATION IN CONDUIT AND TRUNKING: install cables so that they are orderly and capable of being withdrawn.

Arrange single core wiring generally using the loop-in method.

Trunking	:	provide pin racks at 3m intervals in vertical trunking
	:	use cable ties for 3 phase circuits at 2m intervals
	:	use ties as specified
Conduit	:	provide cable clamps in conduit boxes at 10m intervals in vertical conduit
	:	leave sufficient extra length of cable at building construction movement joints for full range of movement

CABLE SURFACE INSTALLATION: dress cables flat, free from twists, kinks and strain, and align parallel to building elements.

When glands and clamps are not required take sheathing of cables into accessory boxes and equipment and sharp edge protection.

CABLE INSTALLATION – FLEXIBLE CORDS: grip cords securely at connections. Where they do not form an integral part of the connected accessory or equipment, provide separate proprietary cord grips. Cables run in ceiling spaces and roof voids shall be run parallel with or at right angles to the beams, steelwork and ceiling joists.

CABLE JOINTING AND TERMINATING GENERALLY: cut all cable ends immediately prior to jointing or terminating. Cables left unconnected for more than 24 hours are to be sealed to permanently prevent the ingress of moisture.

Seal plastic sheathed cables using proprietary shrink on end caps.

For armoured cables clean armour prior to jointing or terminating.

At connections to equipment and switchgear without integral cable clamping terminals, use compression or solder type lugs for bolted terminal connections, of correct bore to fit cable tightly.

For core sizes 10mm^2 and above, form all compression connections to components using tools that cannot be released unless the correct degree of compression has been achieved.

Securely bolt core terminations with lugs to equipment using washers or proprietary shakeproof devices. Connect all cores, including multicore cable space cores, at all joints and terminations. Bond any unused cores of multicore cables to earth at both ends.

CABLE JOINTING AND TERMINATING – ELASTOMER AND PLASTIC INSULATED CABLES: terminate cables using compression glands of the type indicated – complete with:-

	:	earth bond attachment for armoured cables, integral with the body of the gland for cables with conductors larger than 35mm^2
	:	PVC shroud

At core connections to equipment with integral clamping terminals use compression lugs.

CABLE SLEEVES: supply and hand to others for installation non ferrous cable sleeves for incorporation into the structure where cables pass through floors and walls. Pack sleeves with fire resistant material after cable installation.

IDENTIFICATION OF CABLES: shall comply in all respects with Clause 524 of IEE Regulations, and as specified.

SWA AND PVC SHEATHED CABLES:

CABLE TYPES: unless otherwise specified, XLPE insulated PVC sheathed power cables shall have copper conductors, steel wire armouring the same as that for a separate conductor, as stated in the IEE Regulations for Electrical Installations. Single core armoured cables shall have aluminium wire armouring with PVC oversheath. Unless otherwise specified XPLE/ SWA/ PVC cables shall only be used for outdoor applications.

JOINTING OR MAKING OFF TERMINATIONS FOR THERMOPLASTIC INSULATED CABLES: each cable gland shall comply with BS 6121 and be fitted in accordance with the manufacturer's recommendations.

JOINTING OR MAKING OFF TERMINATIONS FOR CONTROL AND INSTRUMENTS CABLES (OTHER THAN TELEPHONE TYPES): the method of stripping the sheath and the insulation shall be as recommended by the cable manufacturer and shall be such that no damage is caused to the insulation or conductor. The cores of each cable shall be identified by ringing through. Numbered ferrules or sleeves or approved crimping tags shall be fitted. The tails shall be neatly arranged and securely laced with PVC cord or approved equivalent and connected in accordance with the diagrams. All spare cores shall be terminated in accordance with the foregoing and, where no terminals are available, the tails shall be clearly exposed and the terminal tags PVC taped. Each spare core shall be left long enough to reach any of the terminals.

MICC INSTALLATION: the runs of cable shall, wherever possible, be concealed behind plaster finish of walls, in ceiling spaces, hollow columns, cast in situ concrete roofs and where provided the maximum use shall be made of floor, vertical and ceiling ducts. All runs shall be made straight and parallel with the sides of the building and all risers and drops shall be vertical. Where cables are run in floor ducts they shall be either fixed to the sides of the duct as for fair faced brickwork or alternatively a timber batten may be first fixed to the duct side and the cables clipped to this. Only on fair faced brickwork and on the underside of a ceiling having no void shall the cables be run on the surface, unless the direct instruction of the Contract Administrator is given to run on finished surfaces.

The fixing of cables in the various locations shall be carried out as follows:

(a) on the surface of unplastered walls and ceilings and in damp situations: heavy gauge one hole fixing spacing saddles and spacing type 'P' clips

(b) on finished surfaces (where permitted): heavy gauge copper saddles

(c) concealed cables: standard copper clips and saddles

Where necessary and with the approval of the Contract Administrator the Contractor may in concealed situations use straps and saddles of special design for the appropriate purpose, made of heavy gauge copper strip.

All saddles, clips and straps shall be secured with brass wood screws, countersunk for one hole fixing saddles, and round headed for all other situations.

The spacing of fixings shall be as follows:

    (a)   cables sunk in floor and roof screeds – 1500mm

    (b)   all other concealed work – 900mm

    (c)   on surface work 250mm

Protection, jointing and sealing as specified.

MICC TERMINATIONS: where cables enter boxes and equipment the following methods of termination shall be adopted.

    (a)   The universal ring type gland (URT) shall be used as standard in all situations other than those specified below and on no occasion shall any other gland be used for main and sub-main cables

    Where the entry to the equipment or box is already tapped the gland shall be screwed direct into the equipment utilising where necessary solid brass reducing sockets

    Where the entry to the equipment or box is a clearance hole the gland shall be fixed with solid brass locknuts. Where space is limited within the equipment then a conduit coupling and bush shall be used between the gland and box

    (b)   The Earthing Screw Type (ES) gland may be used where a tapped entry box is concealed behind a plaster wall finish, panel or ceiling space, always provided that such situation is dry and that both earth screws are tightened

    (c)   Where boxes with lug grip entry are specified for accessories the use of a gland may be dispensed with, the screw on type pot seal shall be clamped into the normal conduit lug grip.

    All spare ways in boxes shall be fitted with brass stopping plugs.

CABLE SUPPORTS:

CABLE CLEATS, TIES, SADDLES AND CLIPS INSTALLATION: for cables on horizontal tray use ties for each circuit. Use tie manufacturer's special tensioning tool where available. Crop off tie ends.

For cables on vertical tray use cleats bolted to tray for paper, plastic or elastomeric insulated cables and saddles or clips for mineral insulated cables.

For cables on vertical or horizontal rack use proprietary fixings to rack for paper, plastic or elastomeric insulated cables and saddles or clips for mineral insulated cables.

Use cleats sized to grip cables firmly without undue pressure or strain on cable, but preventing slipping. Space cleats, ties, saddles and clips to IEE Wiring Regulations.

SECTION H

MEASUREMENT AND PRICING PREAMBLES

FORMAT OF DESCRIPTIONS

In addition to common abbreviations the following have been adopted:-

m	-	Metre
mm	-	Millimetre
m^2	-	Square metre
mm^2	-	Square millimetre
nr	-	number
N	-	Newton
t	-	Tonne
l	-	Litre
C	-	degrees Celsius
BS	-	British Standard
CP	-	Code of Practice

Other metric symbols are given in accordance with BS 6430.

Descriptions are usually given in the plural but secondary phrases which qualify or describe work in addition to the main part of a description may be in the singular.

Every description is to be read as if the phrase "and the like" were incorporated in it.

In order to avoid future amendments, these Preambles may contain references to items which may not necessarily be represented by rates in the Schedule.

The preambles contained in this Section are for guidance only to indicate the basis on which the level of pricing has been made.

All standards referred to within these documents shall be held to be the latest edition published at the date of tender.

A reference to any Act of Parliament or to any Order, Regulation, Statutory Instrument, Building Standard, Code of Practice or the like shall include a reference to any amendment or re-enactment of the same.

DEFINITIONS

"Approved", "Directed", "Selected" and similar expressions shall relate to the decision of the Contract Administrator.

"Works by Others", "Builders work", "Prepared by Others" and similar expressions relate to work done before or to be done during the current contract by Contractors, Sub-Contractors and public bodies directly employed by the Employer.

"Existing": in existence prior to the current contract.

PRICES ALSO TO INCLUDE

"Prices also to include" items which are to be included in the Contractor's Percentage adjustment fall into two categories:-

a) those which appear under the headings "Prices also to include" for which the Contractor is to allow in his Percentage adjustment

b) those which are deemed to be included in descriptions by the method of measurement and are not therefore specifically mentioned but which the Contractor is also obliged to include in his Percentage adjustment.

"Protect the work" shall mean temporarily casing up, covering, protecting and the like to ensure that the work is left clean and perfect at the completion of the works.

"Cut" to include drilling or executing the labour described in any other way.

"Fix Only" shall mean receiving goods and materials, transporting to site, unloading, hoisting, unpacking, assembling, storing, site distribution and positioning, returning packing materials as applicable to the supplier, carriage paid, and obtaining credits thereon in addition to fixing to any backgrounds and connection.

"Supply Only" shall mean the nett invoice costs of goods and materials including carriage less the defrayment of all trade and preferential discounts and credits for packing materials returned to the supplier.

"Connect Only" shall include dismantling, connection and re-assembling equipment supplied and fixed by others.

All fixings are to include their penetration of any intervening soft materials.

SCAFFOLDING

NOTE: the scaffolding rates in this section allow for basic forms of scaffolding. Where more complex forms of scaffolding are required, sub-contract rates should be used.

Scaffolding will only be paid for in accordance with this section when the height of the working platform exceeds 1.50m above ground level. Independent and putlog scaffolds are to be measured at the perimeter of the structure to be scaffolded x the height of the scaffold to comply with the current appropriate regulations.

The rates in this section allow in each item for the cost of scaffolding from ground level to the working platform including boards and double guard rails at one level only.

ELECTRICAL SERVICES

DEFINITIONS

"Fixed" includes fixing with nails, screws, plugs, shot firing pins, bolts, ragbolts, expansion bolts, self drilling anchors, nuts, spring nuts, self locating nuts etc.

"Pipes" includes pipes and tubes

"Bends" includes elbows, easy bends and slow bends both manufactured and made on site of varying degrees and radii.

"Reducers" includes excentric and concentric patterns.

"Brackets" includes standard supports (e.g. clips, saddles, rings, holderbats, hangers) together with component parts.

"Connector" includes any thimble, ferrule, one caulking bush, cap and lining or other fittings or adaptor whether equal or reducing or a combination of any of the above necessary to achieve the connection described.

PRICES ALSO TO INCLUDE

Generally

Fixings: all fixings shall be deemed to be included in the rates regardless of the nature of the background.

Conduit: shall be deemed to include bending, cutting joints and all fittings, fixings and the like, other than those fittings enumerated separately. Draw wires to be included. Conduit is measured excluding fittings.

Conduit Boxes: where these enumerated separately, cutting and jointed conduit to boxes is deemed to be included.

Trunking and Cable Trays: shall be deemed to include couplings and joints, and where applicable, supports, fixings, and drilling accessories and the like. Bolted lapped joints on trays where required. Trunking and tray is measured excluding fittings.

Fittings, accessories and equipment: all rates shall be deemed to include for all necessary fittings and accessories specific to that rate, unless otherwise stated.

Cables/ Final sub-circuit wiring shall be deemed to include all fixings as necessary, and allow for concealed or exposed runs, in existing conduit/ trunking or in new conduit/ trunking.

Terminations for cabling shall be deemed to include all necessary glands, tags, shrouds and the like as necessary. Terminations to equipment will be as described in the specification and is deemed to include all accessories as necessary.

Light fittings shall be deemed to include for all fixings, installation, lamps storage and handling.

Dismantling: all rates where item is removed shall be deemed to include for disconnection and making safe, reconnection of existing circuit, and testing, as necessary.

NOTE

The rates for cable tray and trunking are based on fixing to timber or masonry backgrounds.
The rates do not include for drop rods, unistrut, angle iron or multi trapeze brackets.

ELECTRICAL

PART 2

Global Schedule of Rates

The Global Schedule of Rates computer system can be used on Measured Term Contracts to which any of the NSR Schedule of Rates apply or any other Schedule of Rates. Addendum Schedules are easily added.

For a Works Order or Estimate, the Schedule is either searched on screen and description selected or the required code number is typed in, followed by full dimensions or quantity. Non Schedule codes, such as Star Rates, Net Rates, Dayworks, Invoices, etc., are easily added to a measure. When the completed Order is displayed, dimensions are squared, the resultant quantity is multiplied by the rate and a total Order value generated. Relevant percentages are automatically applied including the Contractor's percentage.

Many reports are available enabling tight control of the contract. These include;
Orders issued but not completed.
Orders issued, completed or billed within a given period.
Orders processed for a selected Building.
Orders completed either within or outside a defined period.
Interim Payments.
How often a particular Nsr code or section of codes has been used on a Contract.
Order printout using, instead of total rate, plant, labour or material rates.

Order details can also be exported to Microsoft Excel and Word or Emailed to another office.

The software will run on any of the Full versions of Windows.

For further information;

Email: john@barcellos.co.uk

Telephone: 0116 233 5559   Fax: 0116 233 5560

Address: Barcellos Limited, Sandbach House, 8 Salisbury Road, Leicester, LE1 7QR

Website: www.barcellos.co.uk

GENERAL INDEX (PART 2)

	Page
Index	2/1
Summary of amendments	2/5
Compilers Notes	2/6
Schedule of Descriptions and Prices	2/17
Basic Prices: Labour, Plant and Materials	BP/1

Save money and complete your financial management jigsaw

ABOUT US

For over 12 years NSR Management have been promoting and developing the National Schedule of Rates on behalf of the Co-authors, the Society of Construction and Quantity Surveyors and the Construction Confederation.

The National Schedule of Rates have become synonymous with the effective management of Measured Term Contracts, thus enabling organisations to be more efficient and save money.

THE NATIONAL SCHEDULE OF RATES

The National Schedule of Rates are in use throughout the whole of the UK, helping hundreds of organisations in both the public and private sector manage and maintain their diverse property needs.

They have also become an invaluable tool for benchmarking and ensuring that planned maintenance agreements are operated as cost effectively as possible.

NSR Management offer a whole range of National Schedule of Rates in book, data and internet subscription format.

The Schedules comprise, in total, approximately 16,000 items of work, covering Housing Maintenance, Building, Mechanical, Electrical, Access Audit, Roadworks and Painting and Decorating.

FINANCIAL BENEFITS FOR YOU AND YOUR ORGANISATION

- **Keeping Costs Down**   More competitive tenders are obtainable from contractors as they become more familiar with the schedules and the level of pricing within them.
- **Information Up Front**   By instructing contractors to use our schedules you will have the benefit of 'up front' cost estimation.
- **Piece of Mind**   The schedules are in use throughout the whole of the UK. and are updated annually to ensure the information is a useful as possible.
- **Good Benchmarking**   The use of our schedules is extensive, hence they are an ideal benchmarking **tool**.

AND FINALLY

NSR Management offer training in the use of the National Schedule of Rates and our 'One Stop Shop' consultancy service is available to assist with contract advice, benchmarking, QS services, condition surveys, contract management and DDA surveys.

Our after sales service is second to none, well that's what our customers tell us, so technical support is on hand for all of our schedules, or any of their associated computer applications.

After all *'Our Business is Your Property'*.

Please contact us for more information

NSR Management Limited
Pembroke Court
22-28 Cambridge Street
Aylesbury, Bucks HP20 1RS

tel: 01296 339966
fax: 01296 338514

nsr@nsrmanagement.co.uk

www.nsrm.co.uk

Index

Listed below are the main Section Titles and Library Groups which have been used in the formulation of the Schedule

Some Section Titles shown do not yet appear in the Schedule, however allocation has been made for their introduction in the future – these are indicated with an asterisk.

NSR ELECTRICAL CODING

NSR Coding A

44	**Contractors General Cost Items**

NSR Coding C

C86	**General lighting and power**

NSR Coding N

25	**Power Installation**
2500	Ancillaries
2550	Connections to Equipment Supplied and Fixed by Others

NSR Coding P

31	**Builders Work: Chases and Holes**

NSR Coding T

37	**Electric Heating Installation**
3700	Water Heaters
3755	Electric Showers
3770	Immersion Heaters

NSR Coding U

40	**Extract Ventilation System**
4000	Extract Fans
4054	Fan Controllers

NSR Coding V

*05	**HV Switchgear**
10	LV Switchgear

1000	Rewireable Fuses
1002	HRC Fuses
1008	MCB's
1012	MCCB's
1016	MCB/RCD's
1018	RCD's
1028	Isolators
1038	Fused Switches
1048	Switched Fuse
1058	Consumer Units
1074	Split Load Consumer Units
15	**Lighting Installation : Luminaires**
1500	Tungsten Luminaires
1524	Fluorescent Luminaires
1545	Discharge Luminaires
1563	Emergency Luminaires
17	**Lighting Installation : Ancillaries and Equipment**
1700	Lighting Switches
1760	Lighting Contactors
1780	Lighting Controls
25	**Power Installation**
2500	DP Switches
2510	Cooker Controls
2520	Ceiling Switches
2525	Socket Outlets
2545	Fused Connections
2570	Cable Outlets

2580	Blanking Plates
2585	Plugs
2590	Back Boxes
30	**Mechanical Services Wiring and Controls**
3000	Isolators
3020	Motor Control Switches
3040	Connections to Equipment Supplied and Fixed by Others
3060	Time Switches
3070	General Fix
32	**Motor Control and Contactor Systems**
3200	Starters
3220	Motor Control Switches
3232	Stop Buttons
3248	Control Panels
3260	Contactors
3280	Motors
37	**Electrical Heating**
3700	Tubular Heating
3706	Hand Driers
3724	Heat Emitters
3760	Thermostats
70	**Earthing and Bonding**
7000	Ancillaries (Bonding Clamps)
7025	Inspection Covers
7050	Earthing Rods
71	**Lightning Protection**
7100	Conductor Tape

7125	Air Terminals
7150	Test and Junction Clamps
85	**System of Wiring**
8500	Conduits
8540	Trunking
8548	Adaptable Boxes
8556	Channel
8568	Cable Tray
86	**System of Wiring : Cables and Ancilliaries**
8700	Electrical Connections
8708	Cables and Conductors
8764	Ancillaries
96	**Sundry Items**
9600	Testing

NSR Coding W

50	**Fire Alarm System**
5000	Batteries
5021	Battery Chargers
5037	Cables and Conductors
5042	Indicator Panels
5063	Ancillaries
5077	Break Glass Contacts
5084	Bells
60	**Security**
6000	Security

65	**Communications/Data**
6500	Data Cable Management
6515	Flood Wiring

SUMMARY OF AMENDMENTS

The alterations to the Schedule are as follows:

1. Section A, contractors general cost items has been amended to more accurately reflect the provisions of SMM7. A99 has now been changed to A44 and Builders work: chases and holes and been moved to section P31 & P32

2. A new section C has been added for refurbishment work where more general stripping out of wiring, conduits etc is required

3. Additional items have also been added as per the building schedule to cover daily charges for scaffold towers, scissor lift, security lighting and alarms

4. A wider range of sangamo timers have been added as items V3062 – 3068

5. Additional items to provide a wider range of hand driers have been added to sections V3712 and V3718

6. The following sections have been added to allow for LSOH wiring:
   V8706011 – 026
   V8710310 – 326
   V8718048 – 062

7. The following items have been deleted as they have been discontinued:
   V8548947, V8548959, V8548967
   V8552945, V8552949, V8552951, V8552959, V8552967, V8552975

COMPILER'S NOTES

The National Schedule for Electrical Services is principally designed for use in the Public Sector, but is now being widely used in the private sector on work with a maintenance or refurbishment bias.

NOTE: The Electrical and Plumbing Services contained in the Building NSR are for works comprising small domestic repairs and modifications forming a minor part of the building works.

The rates contained in the Mechanical and Electrical Schedules are for direct engineering services works of a more substantial nature.

USERS ARE REMINDED THAT THE NATIONAL SCHEDULE IS NOT A PRICING BOOK; FURTHER, THEY ARE ADVISED TO TAKE A SELECTION OF RATES IN THE NATIONAL SCHEDULE AND TEST THEM AGAINST THEIR OWN DATA ON A REGULAR BASIS.

Effective Date of Schedule

The Schedule is revised each year to include latest promulgated wage rates and prices for materials and plant

User's attention is drawn to base dates for labour and materials given in the Compiler's notes and to the listed basic prices.

Users are advised that the rates in this Schedule are deemed to be effective as from 1st August 2008.

CALCULATION OF LABOUR RATES

Plumbers

The productive hours worked by plumbers have been taken as for the Building Operatives.

Hour worked in full year	1802	1802
Less inclement weather 2%	36	
Productive hours	1766	
Non-productive overtime		75
Public holidays (8 x 7.8 hours)		62.5
Hours for which payment is made		1940

	Advanced Plumber £	Trained Plumber £
Basic rate per hour	12.12	10.39
Additional payments for responsibility	0.44	N/A
Sub-total	12.56	10.39
Add bonus 30%	3.768	3.117
Total Cost per hour	**£16.328**	**£13.507**

Plumbers

Build-up of all-in labour rates per annum

			Advanced Plumber £	Trained Plumber £
Earnings	1940	x £16.328	31,678.32	-
Hours		x £13.507	-	26,203.58
Sub-total (A)			31,678.32	26,203.58
National Insurance: secondary contribution (Employer) (rates as from April 2006)				
0% on first £97.01 per week			-	-
12.8% on remainder	12.8% on £26,638.80		3,408.37	
	12.8% on £21,159.06			2,708.36
Industry pension 6.5% of (A)			2,058.96	1,703.23
Annual holiday credit and sickness benefit stamp				
52 weeks	x £35.10		1,825.20	
	x £31.21			1,622.92
Sub-total			38,969.35	32,238.09
Allowance for severance pay		2%)	779.39	644.76
Employer's liability		2%)	779.39	644.76
Trade supervision		6%	2,338.16	1,934.29
Total cost per annum		£	42,866.29	£ 35,461.90
Rate per hour (divide by productive hours 1766)			£23.30	£20.08

Electricians

The hours worked by Electricians are based on a standard working week without any overtime, as follows:

Hour worked in full year	1732.50	17.32.50
Less inclement weather 2%	34.65	
Productive hours	1697.85	
Public holidays (8 x 7.5 hours)		60.00
Hours for which payment is made		1792.50

	Electrician £	Electrician's Labourer £
Basic rate per hour including London addition	13.08	9.67
Add bonus 30%	3.924	2.90
Total Cost per hour	£ 17.004	£12.57

Electricians

Build-up of all-in labour rates per annum

			Electrician £	Electrician's Labourer £
Earnings	1792.5 hrs	x £ 17.004	30,479.67	-
		x £ 12.57	-	22,533.52
National Insurance: secondary contribution (Employer) (rates as from April 2006)				
0% on first £97.01 per week			-	-
12.8% on remainder		12.8% on £25,435.15	3,255.70	
		12.8% on £17,485.88		2,238.59
Industry Pension 6.5% of earnings			1,981.18	1,464.68
Combined benefit scheme 52 weeks				-
wks 3,4,5&6 £75, wks 7-28 £150	Elect		3,450.00	
wks 3,4,5&6 £70, wks 7-28 £140	Lab		-	3,220.00
Total			39,166.55	29,456.79
Allowance for severance pay		2%	783.33	589.14
Employer's liability		2%	783.33	589.14
Trade supervision		6%	2,349.99	1,767.41
Total cost per annum			£ 43,083.20	£ 32,402.47
Rate per hour (Divide by productive hours 1698)			£ 25.38	£ 19.08

©NSR 2008 – 2009

MATERIAL COSTS

The costs of materials are based where possible on Luckins published price lists current at March 2008. It is emphasised that the prices thus obtained are indicative only and the user should satisfy himself as to their validity.

No account has been taken of discounts available under annual purchasing agreements, apart from PVC/SWA/PVC cable where a 90% discount has been applied to reduce distortion

The prices are based on products marketed by various well-known medium priced manufacturers.

QUANTITIES, RATES, EXTENSIONS AND TOTALS

All quantities, rates, extensions and totals are to be kept to two decimal places.

DESCRIPTIONS

Definitions of the methods of measurement used are contained within Part 1, to complement the descriptions within the schedule itself.

COMPUTER-BASED LIBRARY

The schedule is fully computer based. This means that items of work are produced by using the computer to generate descriptions from a custom-compiled Library placed in its memory. This Library is divided into coded sections.

The benefits from using a Library are:

>the codes are identifiable in the Schedule : coding is permanent, compact and does not change with amendments or updates to the Schedule.

>several possibilities for amending the Schedule are available in order to add data, change existing data, etc.

COMPILER'S NOTES – APPENDIX A

COMPUTER-BASED LIBRARY

The Library for the Electrical Schedules is divided into SMM7 (A, N, P, T, U, V and W) main Work Groups, as follows:

A : Contractors General Cost Items

C: Demolition/Alteration/Renovation

N: Power Installation

P: Building Fabric Sundries

T : Electric Heating Installation

U : Extract Ventilation Installation

V : LV Switchgear and Distribution

W : Fire Alarm System

Each of these is divided into a number of Library Groups, for example:

A44 : Contractors General Cost Items

N25 : Power Installation

P31 : Holes/chases/covers/supports for plumbing or mechanical services

T37 : Electric Heating Installation

U 40 : Extract Ventilation Installation

V10 : LV Switchgear and Distribution

W50 : Fire Alarm System

Each of these is then divided into Combined Headings, for example:

T3700 : Water Heaters : remove damaged item of equipment, supply and fix new item of equipment

T3755 : Electric Showers ; fix only instantaneous electric shower, flexible hose, rail, shower head, or swivel spout

These are then divided into Work Heads (item descriptions) as follows:

020   :   Heating element

022   :   Thermostat

An example of the way in which these codes and texts appear in the Schedule is as follows:

T : ELECTRICAL SERVICES : WORKS OF ALTERATION/SMALL WORKS/REPAIRS.

T37 : Electric Heating Installation (Library Group T37)

T3700 : Water Heaters : remove damaged item of equipment, supply and Fix new item of equipment   (Combined heading T3700)

020 : Heating element (Work Head 020)

022 : Thermostat (Work Head 022)

T3755 : Electric Showers ; fix only instantaneous electric shower, flexible hose, rail, shower head, or swivel spout

050 : 7 – 10 kW Domestic type

052 : 7 – 10 kW Industrial type

The Library contains a number of headings and items that for compactness it has been decided not to call up for use in the Schedule. These provide scope for future amendments to the Schedule.

Several possibilities for future amendment to the schedule are available:

- additional coding can produce an item not currently printed in the Schedule by a combination of Combined Headings (CH) and Work Heads (WH) already in the Library. Such an item could be said to be in the form

    existing Library Group/existing CH/<u>new</u> WH

    existing Library Group/<u>new</u> CH/existing WH

    existing Library Group/<u>new</u> CH/<u>new</u> WH

  a <u>new</u> Library Group could use any of the three foregoing combinations :

in every case a unique code will be produced so that the required new item goes into its proper place in the Schedule and retains its number permanently.

Users introducing their own descriptions into the Schedule for their own use, are advised not to use the omitted numbers in the published Schedule since these "missing" numbers could lead to an obvious conflict but can easily be avoided if users introduce their own references in the form of suffix letters.

e.g.   T3755/050A to define a new item (unique to User) which requires to be entered into the Schedule by the User immediately following the standard published item coded T3755/050.

Diagnostics

Initial fault finding for repairs of a simple nature is deemed to be included in the rates i.e. those faults which would be identified within a few minutes by a competent tradesman. Inability to identify reason for a fault within say, fifteen minutes should be reported to the client and further investigation may be paid for on dayworks if the client agrees that the difficulty of diagnosis warrants it. Servicing, testing and inspection are not currently shown as separate items, however these can be added separately to the Schedule at tender stage.

Testing – Electrical Installations

The post-installation testing to ensure compliance with IEE regulations, Electricity at Work Regulations and NIC guidelines, is limited to the work ordered and executed. The electrician is to satisfy himself that the circuit is safe, i.e. carry out continuity and polarity checks, which are included in the rates. Further testing such as earth loop impedence, insulation resistance or under sized earthing are not included in the rates. The reimbursement for these costs are to be part of the percentage adjustment. However, the contractor is expected to report to the client, any failures to comply, which are apparent without testing of the existing equipment and installation on which no work has been ordered. These reports should be in writing but would be limited to simple observations of the existing system made by the tradesman in the area where work has been carried out. The rates do not include for the NIC periodic inspections, as these are deemed to be part of a maintenance inspection programme.

Specification

The specification included in Part 1 only provides general materials and workmanship preambles. It is assumed that generally the client will issue a schedule of preferred suppliers and manufacturers for the contract, as part of a supplementary equipment specification, which the tenderer will take into account when assessing his percentage adjustment.

The manufacturers described in the measurement and pricing preambles are provided only as a guide to the source of prices shown and therefore for information only. The tenderer is to assess the level of prices indicated against the stipulations of the clients specification or the requirements of the site and adjust his percentage accordingly.

Specialist items required by the client are to be listed in the enquiry to tenderers i.e.:

Halon Installation
Controls and BMS
Security Systems

N.B.

1. Cables are priced from rolls of either 50 or 100 metre rolls and no allowance has been made for cutting or for cut lengths.

2. Many components are available in specified lengths and no allowance has been made for the cutting to the schedules measured unit of measure i.e. metre. (e.g. conduit is available in 3 metre lengths).

3. Where materials of a specified size are not available, alternative materials are substituted and the alteration to the schedule has been carried out.

4. Pricing policies of the various manufacturers used, especially cable manufacturers, distort the real cost of the materials by the various discounts against the different materials. Generally no allowance has been made for these discounts, however, due to comments from users, an allowance has been made in the majority of the wiring items in order to eliminate the anomaly.

7. Where there are descriptions in the schedule which have materials that are no longer available or the materials have been superseded, an indicative cost has been assigned as possible sources might still be available, but not listed.

8. Where items in the schedule which refer to BS numbers which have been superceded, the item is to be in accordance with the new number e.g. BS2871 should be read as BSEN1057

ELECTRICAL

SCHEDULE OF DESCRIPTIONS AND PRICES

One STOP Shop

Our Business is your Property

On Site Training in the use of National Schedules

Our standard full day training course 9.30am – 4.30pm comprises the following sessions:

- Benefits of Measured Term Contract
- Estimate Percentage 'A'
- JCT Measured Term Contract
- Communication
- Dispute Resolution
- Some Practical Problems
- Estimating the Value of Works Orders

We usually suggest that numbers should be restricted to no more than 8 and the total cost for the training is £1250 [+VAT] plus speaker's expenses which should not exceed £200. If more that 8 attend the cost for each additional delegate is £200

You could always invite others from out with your company to defray costs if you have insufficient in house staff who wish to participate.

Shorter sessions can be organised to suit your particular needs and the cost would be adjusted accordingly.

CONTRACT ADVICE BENCHMARKING SERVICES SCHEDULE OF RATES QS SERVICES

CONDITION SURVEYS MANAGEMENT OF CONTRACTS DDA SURVEYS

NSR Management Ltd
Pembroke Court, 22-28 Cambridge Street, Aylesbury, HP20 1RS Telephone: 01296 339966 Facsimile: 01296 338514
e-mail: nsrm@nsrmanagement.co.uk www.nsrm.co.uk Est. 1995 Registered in England 3574827 VAT No. 640 0550 83

			Mat. £	Lab. £	Plant £	Total £
	A : ELECTRICAL SERVICES : WORKS OF ALTERATION/SMALL WORKS					
	A44 : CONTRACTORS GENERAL COST ITEMS					
A4410	**Erect and dismantle scaffolding**					
050	Independent tied scaffold	m2		9.50		9.50
060	Putlog scaffold	m2		7.49		7.49
070	Additional boarded platform (measured on plan)	m2		7.49		7.49
080	Brick guards	nr		1.15		1.15
085	Fans	m2		8.64		8.64
087	Roof edge double guardrail	m		6.34		6.34
090	Temporary roofs (Plastic tarpaulin covered)	m2		12.96		12.96
092	Temporary dust screens (Plastic tarpaulin)	m2		3.46		3.46
094	Rubbish chute	m		1.44		1.44
096	Scaffold Hoist	nr		4.32		4.32
A4415	**Erect and dismantle chimney scaffold ne 2.00 m girth**					
100	1.50 - 3.00 m From ground level	nr		43.20		43.20
105	3.00 - 4.00 m From ground level	nr		57.60		57.60
110	4.00 - 5.00 m From ground level	nr		72.00		72.00
115	5.00 - 6.00 m From ground level	nr		86.40		86.40
A4420	**Erect and dismantle chimney scaffold 2.00 - 3.00 girth**					
100	1.50 - 3.00 m From ground level	nr		64.80		64.80
105	3.00 - 4.00 m From ground level	nr		86.40		86.40
110	4.00 - 5.00 m From ground level	nr		108.00		108.00
115	5.00 - 6.00 m From ground level	nr		129.60		129.60
A4425	**Erect and dismantle tower scaffold to provide a working platform of 2.50 x 0.85 m**					
100	1.50 - 3.00 m From ground level	nr		21.60		21.60
105	3.00 - 4.00 m From ground level	nr		25.92		25.92
110	4.00 - 5.00 m From ground level	nr		29.38		29.38
115	5.00 - 6.00 m From ground level	nr		34.56		34.56

© NSR 01 Aug 2008 - 31 Jul 2009   A1

	A : ELECTRICAL SERVICES : WORKS OF ALTERATION/SMALL WORKS		Mat. £	Lab. £	Plant £	Total £
A4430	Erect and dismantle tower scaffold to provide a working platform of 2.50 x 1.45 m					
100	1.50 - 3.00 m From ground level	nr		23.62		23.62
105	3.00 - 4.00 m From ground level	nr		28.51		28.51
110	4.00 - 5.00 m From ground level	nr		32.26		32.26
115	5.00 - 6.00 m From ground level	nr		38.02		38.02
A4435	Erect and dismantle lightweight aluminium access units					
200	Chimney scaffold unit to provide working platform to half of centre ridge stack	nr		8.64		8.64
205	Chimney scaffold unit to provide complete working platform around centre ridge stack	nr		17.28		17.28
230	Window access unit, 450 mm wide platform	nr		4.32		4.32
235	Window access unit, 600 mm wide platform	nr		5.18		5.18
240	Staircase access unit, 300 - 450 mm wide platform	nr		4.32		4.32
245	Staircase access unit, 600 - 675 mm wide platform	nr		5.18		5.18
A4440	Partially and temporarily dismantle and re-erect scaffolding for safety reasons					
050	Independent tied scaffold	m2		7.20		7.20
060	Putlog scaffold	m2		7.49		7.49
070	Additional boarded platform (measured on plan)	m2		7.20		7.20
080	Brick guards	nr		1.15		1.15
085	Fans	m2		8.64		8.64
090	Temporary roofs (Plastic tarpaulin covered)	m2		12.96		12.96
A4445	Partially and temporarily dismantle and re-erect chimney scaffold ne 2.00 m girth for safety reasons					
100	1.50 - 3.00 m From ground level	nr		36.00		36.00
105	3.00 - 4.00 m From ground level	nr		48.96		48.96
110	4.00 - 5.00 m From ground level	nr		61.92		61.92
115	5.00 - 6.00 m From ground level	nr		73.44		73.44
A4450	Partially and temporarily dismantle and re-erect chimney scaffold 2.00 - 3.00 m girth for safety reasons					
100	1.50 - 3.00 m From ground level	nr		54.72		54.72
105	3.00 - 4.00 m From ground level	nr		73.44		73.44
110	4.00 - 5.00 m From ground level	nr		92.16		92.16
115	5.00 - 6.00 m From ground level	nr		109.44		109.44

© NSR 01 Aug 2008 - 31 Jul 2009

A2

A : ELECTRICAL SERVICES : WORKS OF ALTERATION/SMALL WORKS				Mat. £	Lab. £	Plant £	Total £
A4455	Partially and temporarily dismantle and re-erect tower scaffold to provide a working platform of 2.50 x 0.85 m for safety reasons						
100	1.50 - 3.00 m From ground level		nr		18.14		18.14
105	3.00 - 4.00 m From ground level		nr		21.60		21.60
110	4.00 - 5.00 m From ground level		nr		24.77		24.77
115	5.00 - 6.00 m From ground level		nr		29.38		29.38
A4460	Partially and temporarily dismantle and re-erect tower scaffold to provide a working platform of 2.50 x 1.45 m for safety reasons						
100	1.50 - 3.00 m From ground level		nr		20.16		20.16
105	3.00 - 4.00 m From ground level		nr		24.19		24.19
110	4.00 - 5.00 m From ground level		nr		27.36		27.36
115	5.00 - 6.00 m From ground level		nr		32.26		32.26
A4462	Erect and dismantle temporary security fencing						
300	3500 x 2200 mm galvanised mesh panels, placed in precast concrete base feet, fixing with clamps		nr		2.80		2.80
A4465	Scaffolding - Hire charge (weekly rate)						
050	Independent tied scaffold		m2			3.58	3.58
060	Putlog scaffold		m2			2.71	2.71
070	Additional boarded platform (measured on plan)		m2			3.14	3.14
080	Brick guards		nr			0.62	0.62
085	Fans		m2			4.66	4.66
087	Roof edge double guardrail		m			2.53	2.53
090	Temporary roofs (Plastic tarpaulin covered)		m2			0.90	0.90
092	Temporary dust screens (Plastic tarpaulin)		m2			0.90	0.90
094	Rubbish chute		m2			12.75	12.75
096	Scaffold hoist		nr			120.00	120.00
A4470	Chimney scaffold ne 2.00 m girth - Hire charge (weekly rate)						
100	1.50 - 3.00 m From ground level		nr			73.36	73.36
105	3.00 - 4.00 m From ground level		nr			100.32	100.32
110	4.00 - 5.00 m From ground level		nr			127.29	127.29
115	5.00 - 6.00 m From ground level		nr			154.25	154.25
A4475	Chimney scaffold 2.00 - 3.00 m girth - Hire charge (weekly rate)						
100	1.50 - 3.00 m From ground level		nr			111.56	111.56

© NSR 01 Aug 2008 - 31 Jul 2009   A3

A : ELECTRICAL SERVICES : WORKS OF ALTERATION/SMALL WORKS			Mat. £	Lab. £	Plant £	Total £
A4475						
105	3.00 - 4.00 m From ground level	nr			138.52	138.52
110	4.00 - 5.00 m From ground level	nr			165.49	165.49
115	5.00 - 6.00 m From ground level	nr			182.01	182.01
A4480	Tower scaffold to provide a working platform of 2.50 x 0.85 m - Hire charge (Weekly rate)					
100	1.50 - 3.00 m From ground level	nr			113.50	113.50
105	3.00 - 4.00 m From ground level	nr			133.00	133.00
110	4.00 - 5.00 m From ground level	nr			152.50	152.50
115	5.00 - 6.00 m From ground level	nr			172.00	172.00
A4482	Tower scaffold to provide a working platform of 2.50 x 0.85 m - Hire charge Daily rate)					
100	1.50 - 3.00 m From ground level	nr			56.75	56.75
105	3.00 - 4.00 m From ground level	nr			66.50	66.50
110	4.00 - 5.00 m From ground level	nr			76.25	76.25
115	5.00 - 6.00 m From ground level	nr			86.00	86.00
A4485	Tower scaffold to provide a working platform of 2.50 x 1.45 m - Hire charge (Weekly rate)					
100	1.50 - 3.00 m From ground level	nr			113.50	113.50
105	3.00 - 4.00 m From ground level	nr			133.00	133.00
110	4.00 - 5.00 m From ground level	nr			152.50	152.50
115	5.00 - 6.00 m From ground level	nr			172.00	172.00
A4487	Tower scaffold to provide a working platform of 2.50 x 1.45 m - Hire charge (Daily rate)					
100	1.50 - 3.00 m From ground level	nr			56.75	56.75
105	3.00 - 4.00 m From ground level	nr			66.50	66.50
110	4.00 - 5.00 m From ground level	nr			76.25	76.25
115	5.00 - 6.00 m From ground level	nr			86.00	86.00
A4490	Lightweight aluminium access units - Hire charge (weekly rate)					
200	Chimney scaffold unit to provide working platform to half of centre ridge stack	nr			110.00	110.00
205	Chimney scaffold unit to provide complete working platform around centre ridge stack	nr			220.00	220.00

© NSR 01 Aug 2008 - 31 Jul 2009   A4

	A : ELECTRICAL SERVICES : WORKS OF ALTERATION/SMALL WORKS		Mat. £	Lab. £	Plant £	Total £
A4490						
230	Window access unit, 450 mm wide platform	nr			90.00	90.00
235	Window access unit, 600 mm wide platform	nr			90.00	90.00
240	Staircase access unit, 300 - 450 mm wide platform	nr			90.00	90.00
245	Staircase access unit, 600 - 675 mm wide platform	nr			90.00	90.00
A4491	Lightweight aluminium access units - Hire charge (Daily rate)					
200	Chimney scaffold unit to provide working platform to half of centre ridge stack	nr			55.00	55.00
205	Chimney scaffold unit to provide complete working platform around centre ridge stack	nr			110.00	110.00
230	Window access unit, 450 mm wide platform	nr			45.00	45.00
235	Window access unit, 600 mm wide platform	nr			45.00	45.00
240	Staircase access unit, 300 - 450 mm wide platform	nr			45.00	45.00
245	Staircase access unit, 600 - 675 mm wide platform	nr			45.00	45.00
A4492	Temporary security fencing - Hire charges (Weekly rate 1 to 3 weeks)					
300	3500 x 2200 mm galvanised mesh panels, placed in precast concrete base feet, fixing with clamps	nr			7.02	7.02
A4494	Temporary heating/lighting - Hire charge (weekly rate)					
350	Festoon lighting	nr			32.00	32.00
360	Temporary heating (per room)	nr			19.00	19.00
A4495	Mechanical Access equipment - Hire charge (weekly rate)					
300	Compact scissor lift 7.8m	nr			360.00	360.00
A4496	Mechanical Access equipment - Hire charge (Daily rate)					
300	Compact scissor lift 7.8m	nr			216.00	216.00
A4497	Scaffold Lighting/Security - Hire charge (Minimum hire & weekly rate)					
300	Security floor lighting, per light - Set up and minimum charge	nr			50.00	50.00
310	Security floor lighting	nr			3.50	3.50
320	Scaffold alarm, Quad beam and passive 50m range - Set up and minimum charge	nr			234.00	234.00
330	Scaffold alarm, Quad beam and passive 50m range	nr			21.00	21.00

© NSR 01 Aug 2008 - 31 Jul 2009   A5

A : ELECTRICAL SERVICES : WORKS OF ALTERATION/SMALL WORKS	Mat. £	Lab. £	Plant £	Total £

© NSR 01 Aug 2008 - 31 Jul 2009 — A6

			Mat. £	Lab. £	Plant £	Total £
	C : DEMOLITION/ALTERATION/RENOVATION					
	C86 : General lighting and power					
C8610	**Generally**					
051	Stripping out surface steel conduits.	m		3.71		3.71
052	Stripping out surface PVC conduits.	m		2.22		2.22
054	Stripping out surface steel trunking system.	m		4.45		4.45
056	Stripping out surface PVC trunking system.	m		3.71		3.71
061	Stripping out surface cables, isolating.	m		2.22		2.22
062	Stripping out single core cables from conduit.	m		0.74		0.74
064	Stripping out cables from trunking.	m		0.32		0.32
102	Stripping out distribution boxes, isolating.	nr		6.67		6.67
122	Stripping out fuse boxes.	nr		3.71		3.71
142	Stripping out consumer units, isolating.	nr		6.67		6.67
162	Stripping out surface light switches, isolating.	nr		2.22		2.22
202	Stripping out pull cord ceiling switches, isolating.	nr		4.45		4.45
242	Stripping out surface socket outlets, isolating.	nr		2.96		2.96
252	Stripping out flush socket outlets, isolating.	nr		2.22		2.22
272	Stripping out cooker control units, isolating.	nr		2.96		2.96
292	Stripping out lamp holders, isolating.	nr		3.71		3.71
312	Stripping out batten lamp holders, isolating.	nr		4.45		4.45
332	Stripping out ceiling roses and pendants, isolating.	nr		4.45		4.45
347	Stripping out window/wall mounted vent axia fans	nr		22.23		22.23
348	Stripping out electric showers	nr		22.23		22.23

© NSR 01 Aug 2008 - 31 Jul 2009   C1

C : DEMOLITION/ALTERATION/RENOVATION	Mat. £	Lab. £	Plant £	Total £

© NSR 01 Aug 2008 - 31 Jul 2009 — C2

			Mat. £	Lab. £	Plant £	Total £
	N : ELECTRICAL SERVICES : WORKS OF ALTERATION/SMALL WORKS					
	N25 : POWER INSTALLATION					
N2500	**ANCILLARIES : disconnect item of equipment and leave safe**					
013	Cooker, 45 amp	nr		4.23		4.23
N2550	**CONNECTIONS TO EQUIPMENT SUPPLIED AND FIXED BY OTHERS : connect only, ancillary items including final flexible connection**					
010	Motor, fractional - 1 hp	nr		12.69		12.69
012	Motor, 1 - 3 hp	nr		19.04		19.04
014	Motor, 3 - 6 hp	nr		25.38		25.38
016	Bain-marie	nr		25.38		25.38
018	Waste disposal unit	nr		16.92		16.92
020	Deep freeze	nr		12.69		12.69
022	Refrigerator	nr		12.69		12.69
024	Dish-washer	nr		12.69		12.69
026	Spin dryer	nr		12.69		12.69
028	Washing machine	nr		12.69		12.69
030	Cooker extract hood	nr		12.69		12.69
032	Cooker	nr		12.69		12.69

© NSR 01 Aug 2008 - 31 Jul 2009

N1

N : ELECTRICAL SERVICES : WORKS OF ALTERATION/SMALL WORKS	Mat. £	Lab. £	Plant £	Total £

			Mat. £	Lab. £	Plant £	Total £
	P : BUILDING FABRIC SUNDRIES					
	P31 : Holes/chases/covers/supports for plumbing or mechanical services					
P3150	**general builders work**					
300	Cutting or forming holes for pipes ne 55 mm nominal size in concrete 100 mm thick, making good	nr	0.37	3.88	4.56	8.81
302	Cutting or forming holes for pipes ne 55 mm nominal size in concrete 200 mm thick, making good	nr	0.37	4.74	5.96	11.07
304	Cutting or forming holes for pipes ne 55 mm nominal size in common brickwork 102 mm thick, making good	nr	0.56	3.02	3.56	7.14
306	Cutting or forming holes for pipes ne 55 mm nominal size in common brickwork 215 mm thick, making good	nr	0.56	3.23	3.99	7.78
308	Cutting or forming holes for pipes ne 55 mm nominal size in common brickwork 328 mm thick, making good	nr	0.56	4.31	6.01	10.88
310	Cutting or forming holes for pipes ne 55 mm nominal size in facing brickwork 102 mm thick, making good	nr	0.56	3.88	4.56	9.00
312	Cutting or forming holes for pipes ne 55 mm nominal size in facing brickwork 215 mm thick, making good	nr	0.56	4.31	5.21	10.08
314	Cutting or forming holes for pipes ne 55 mm nominal size in facing brickwork 328 mm thick, making good	nr	0.56	5.82	7.59	13.97
316	Cutting or forming holes for pipes ne 55 mm nominal size in blockwork 60 mm thick, making good	nr	0.37	1.29	1.45	3.11
318	Cutting or forming holes for pipes ne 55 mm nominal size in blockwork 75 mm thick, making good	nr	0.37	1.51	1.68	3.56
320	Cutting or forming holes for pipes ne 55 mm nominal size in blockwork 100 mm thick, making good	nr	0.37	1.72	1.96	4.05
322	Cutting or forming holes for pipes ne 55 mm nominal size in blockwork 150 mm thick, making good	nr	0.37	1.72	2.01	4.10
324	Cutting or forming holes for pipes ne 55 mm nominal size in blockwork 200 mm thick, making good	nr	0.37	1.72	2.06	4.15
326	Cutting or forming holes for pipes ne 55 mm nominal size in blockwork 215 mm thick, making good	nr	0.37	1.94	2.28	4.59
350	Cutting or forming holes for pipes 55 to 110 mm nominal size in concrete 100 mm thick, making good	nr	0.75	5.39	6.14	12.28
352	Cutting or forming holes for pipes 55 to 110 mm nominal size in concrete 200 mm thick, making good	nr	1.12	6.90	8.22	16.24
354	Cutting or forming holes for pipes 55 to 110 mm nominal size in common brickwork 102 mm thick, making good	nr	0.75	4.31	4.91	9.97
356	Cutting or forming holes for pipes 55 to 110 mm nominal size in common brickwork 215 mm thick, making good	nr	1.12	5.39	6.24	12.75
358	Cutting or forming holes for pipes 55 to 110 mm nominal size in common brickwork 328 mm thick, making good	nr	1.12	6.90	8.72	16.74

© NSR 01 Aug 2008 - 31 Jul 2009

	P : BUILDING FABRIC SUNDRIES		Mat. £	Lab. £	Plant £	Total £
P3150						
360	Cutting or forming holes for pipes 55 to 110 mm nominal size in facing brickwork 102 mm thick, making good	nr	1.12	5.82	6.59	13.53
362	Cutting or forming holes for pipes 55 to 110 mm nominal size in facing brickwork 215 mm thick, making good	nr	1.12	6.90	7.92	15.94
364	Cutting or forming holes for pipes 55 to 110 mm nominal size in facing brickwork 328 mm thick, making good	nr	1.12	9.48	11.63	22.23
366	Cutting or forming holes for pipes 55 to 110 mm nominal size in blockwork 60 mm thick, making good	nr	0.75	1.51	1.68	3.94
368	Cutting or forming holes for pipes 55 to 110 mm nominal size in blockwork 75 mm thick, making good	nr	0.75	2.59	2.71	6.05
370	Cutting or forming holes for pipes 55 to 110 mm nominal size in blockwork 100 mm thick, making good	nr	0.75	2.59	2.91	6.25
372	Cutting or forming holes for pipes 55 to 110 mm nominal size in blockwork 150 mm thick, making good	nr	0.75	2.80	3.23	6.78
374	Cutting or forming holes for pipes 55 to 110 mm nominal size in blockwork 200 mm thick, making good	nr	0.75	3.23	3.64	7.62
376	Cutting or forming holes for pipes 55 to 110 mm nominal size in blockwork 215 mm thick	nr	0.75	3.23	3.74	7.72
400	Cutting or forming holes for pipes over 110 mm nominal size in concrete 100 mm thick, making good	nr	0.93	7.54	8.40	16.87
402	Cutting or forming holes for pipes over 110 mm nominal size in concrete 200 mm thick, making good	nr	1.68	9.48	11.43	22.59
404	Cutting or forming holes for pipes over 110 mm nominal size in common brickwork 102 mm thick, making good	nr	1.68	6.03	6.72	14.43
406	Cutting or forming holes for pipes over 110 mm nominal size in common brickwork 215 mm thick, making good	nr	1.68	7.33	8.27	17.28
408	Cutting or forming holes for pipes over 110 mm nominal size in common brickwork 328 mm thick, making good	nr	1.68	9.05	10.98	21.71
410	Cutting or forming holes for pipes over 110 mm nominal size in facing brickwork 102 mm thick, making good	nr	1.68	7.97	8.85	18.50
412	Cutting or forming holes for pipes over 110 mm nominal size in facing brickwork 215 mm thick, making good	nr	1.68	9.48	10.63	21.79
414	Cutting or forming holes for pipes over 110 mm nominal size in facing brickwork 328 mm thick, making good	nr	1.68	12.28	14.56	28.52
416	Cutting or forming holes for pipes over 110 mm nominal size in blockwork 60 mm thick, making good	nr	1.12	2.15	2.36	5.63
418	Cutting or forming holes for pipes over 110 mm nominal size in blockwork 75 mm thick, making good	nr	1.12	3.23	3.49	7.84
420	Cutting or forming holes for pipes over 110 mm nominal size in blockwork 100 mm thick, making good	nr	1.12	3.45	3.81	8.38
422	Cutting or forming holes for pipes over 110 mm nominal size in blockwork 150 mm thick, making good	nr	1.12	3.66	4.14	8.92

© NSR 01 Aug 2008 - 31 Jul 2009

	P : BUILDING FABRIC SUNDRIES		Mat. £	Lab. £	Plant £	Total £
P3150						
424	Cutting or forming holes for pipes over 110 mm nominal size in blockwork 200 mm thick, making good	nr	1.12	4.31	4.76	10.19
426	Cutting or forming holes for pipes over 110 mm nominal size in blockwork 215 mm thick, making good	nr	1.12	4.53	4.99	10.64
450	Cutting or forming chases for services 1 nr 19 mm dia in concrete, making good	m	0.19	5.55	0.96	6.70
452	Cutting or forming chases for services 1 nr 19 mm dia in brickwork, making good	m	0.19	4.31	1.86	6.36
454	Cutting or forming chases for services 1 nr 19 mm dia in blockwork, making good	m	0.19	2.15	0.93	3.27
500	Cutting or forming chases for services 1 nr 25 mm dia in concrete, making good	m	0.37	7.40	1.28	9.05
502	Cutting or forming chases for services 1 nr 25 mm dia in brickwork, making good	m	0.37	5.17	1.96	7.50
504	Cutting or forming chases for services 1 nr 25 mm dia in blockwork, making good	m	0.37	2.59	0.98	3.94
506	Cutting or forming chases for services 1 nr 50 mm dia in concrete, making good	m	0.75	13.87	2.40	17.02
508	Cutting or forming chases for services 1 nr 50 mm dia in brickwork, making good	m	0.75	7.33	2.21	10.29
510	Cutting or forming chases for services 1 nr 50 mm dia in blockwork, making good	m	0.75	3.88	1.13	5.76
550	Cutting or forming chases for services 1 nr 50 x 50 mm in concrete, making good	m	0.75	13.87	2.40	17.02
552	Cutting or forming chases for services 1 nr 50 x 50 mm in brickwork, making good	m	0.75	6.46	2.11	9.32
554	Cutting or forming chases for services 1 nr 50 x 50 mm in blockwork, making good	m	0.75	3.45	1.08	5.28
600	Cutting or forming chases for services 2 nr 19 mm dia in concrete, making good	m	0.56	9.71	1.68	11.95
602	Cutting or forming chases for services 2 nr 19 mm dia in brickwork, making good	m	0.56	7.76	2.25	10.57
604	Cutting or forming chases for services 2 nr 19 mm dia in blockwork, making good	m	0.56	3.88	1.13	5.57
606	Cutting or forming chases for services 1 nr 100 x 100 mm in concrete, making good	m	1.86	31.44	5.44	38.74
608	Cutting or forming chases for services 1 nr 100 x 100 mm in brickwork, making good	m	1.86	12.07	2.75	16.68
610	Cutting or forming chases for services 1 nr 100 x 100 mm in blockwork, making good	m	1.86	7.54	1.55	10.95
625	Cutting or forming chases for services 2 nr 25 mm dia in concrete, making good	m	0.75	13.41	2.32	16.48

© NSR 01 Aug 2008 - 31 Jul 2009

	P : BUILDING FABRIC SUNDRIES		Mat. £	Lab. £	Plant £	Total £
P3150						
627	Cutting or forming chases for services 2 nr 25 mm dia in brickwork, making good	m	0.75	9.27	2.43	12.45
629	Cutting or forming chases for services 2 nr 25 mm dia in blockwork, making good	m	0.75	4.53	1.20	6.48
650	Cutting or forming chases for services 2 nr 50 mm dia in concrete, making good	m	1.86	24.97	4.32	31.15
652	Cutting or forming chases for services 2 nr 50 mm dia in brickwork, making good	m	1.86	11.64	2.70	16.20
654	Cutting or forming chases for services 2 nr 50 mm dia in blockwork, making good	m	1.86	6.90	1.48	10.24
656	Cutting or forming chases for services 1 nr 150 x 100 mm in concrete, making good	m	3.73	37.91	6.56	48.20
658	Cutting or forming chases for services 1 nr 150 x 100 mm in brickwork, making good	m	3.73	16.38	2.57	22.68
660	Cutting or forming chases for services 1 nr 150 x 100 mm in blockwork, making good	m	3.73	10.78	1.92	16.43
P3160	**Labours in reinforced concrete where reinforcement cut (50% addition on chases and holes in unreinforced concrete)**					
300	Cutting or forming holes for pipes ne 55 mm nominal size in concrete 100 mm thick, making good	nr	0.37	5.82	6.84	13.03
302	Cutting or forming holes for pipes ne 55 mm nominal size in concrete 200 mm thick, making good	nr	0.37	7.11	8.45	15.93
350	Cutting or forming holes for pipes 55 to 110 mm nominal size in concrete 100 mm thick, making good	nr	0.75	7.97	9.10	17.82
352	Cutting or forming holes for pipes 55 to 110 mm nominal size in concrete 200 mm thick, making good	nr	1.12	10.34	12.33	23.79
400	Cutting or forming holes for pipes over 110 mm nominal size in concrete 100 mm thick, making good	nr	0.93	11.21	12.23	24.37
402	Cutting or forming holes for pipes over 110 mm nominal size in concrete 200 mm thick, making good	nr	1.68	14.22	17.14	33.04
450	Cutting or forming chases for services 1 nr 19 mm dia in concrete, making good	m	0.19	8.32	2.82	11.33
500	Cutting or forming chases for services 1 nr 25 mm dia in concrete, making good	m	0.37	11.10	3.35	14.82
506	Cutting or forming chases for services 1 nr 50 mm dia in concrete, making good	m	0.75	20.80	5.02	26.57
550	Cutting or forming chases for services 1 nr 50 x 50 mm in concrete, making good	m	0.75	20.80	5.02	26.57
600	Cutting or forming chases for services 2 nr 19 mm dia in concrete, making good	m	0.56	14.33	3.88	18.77
606	Cutting or forming chases for services 1 nr 100 x 100 mm in concrete, making good	m	1.86	47.16	9.65	58.67

© NSR 01 Aug 2008 - 31 Jul 2009

	P : BUILDING FABRIC SUNDRIES		Mat. £	Lab. £	Plant £	Total £
P3160						
625	Cutting or forming chases for services 2 nr 25 mm dia in concrete, making good	m	0.75	19.88	4.86	25.49
650	Cutting or forming chases for services 2 nr 50 mm dia in concrete, making good	m	1.86	37.45	7.94	47.25
656	Cutting or forming chases for services 1 nr 150 x 100 mm in concrete, making good	m	3.73	56.87	11.36	71.96
	P32 : Holes/chases/covers/supports for electrical services					
P3270	**General builders work**					
800	Cutting or forming holes, mortices, sinkings and chases, concealed service, luminaire point, making good	nr	0.49	25.17		25.66
802	Cutting or forming holes, mortices, sinkings and chases, concealed service, socket outlet point, making good	nr	0.43	18.13		18.56
804	Cutting or forming holes, mortices, sinkings and chases, concealed service, fitting outlet point, making good	nr	0.43	18.13		18.56
806	Cutting or forming holes, mortices, sinkings and chases, concealed service, equipment and control gear point, making good	nr	0.59	30.21		30.80
850	Cutting or forming holes, mortices, sinkings and chases, exposed service, luminaire point, making good	nr	0.17	14.10		14.27
852	Cutting or forming holes, mortices, sinkings and chases, exposed service, socket outlet point, making good	nr	0.14	9.06		9.20
854	Cutting or forming holes, mortices, sinkings and chases, exposed service, fitting outlet point, making good	nr	0.14	9.06		9.20
856	Cutting or forming holes, mortices, sinkings and chases, exposed service, equipment and control gear point, making good	nr	0.20	15.10		15.30

© NSR 01 Aug 2008 - 31 Jul 2009

P : BUILDING FABRIC SUNDRIES	Mat. £	Lab. £	Plant £	Total £

			Mat. £	Lab. £	Plant £	Total £
	T : ELECTRICAL SERVICES : WORKS OF ALTERATION/SMALL WORKS					
	T37 : ELECTRIC HEATING INSTALLATION					
T3700	WATER HEATERS : remove damaged item of equipment, supply and fix new item of equipment					
020	Heating element	nr	40.00	12.69		52.69
022	Thermostat	nr	11.75	8.46		20.21
T3705	WATER HEATERS : remove damaged water heater, fix only new water heater, over or under sink application, swivel spout, thermostatically controlled					
025	7 Litre, 3kW, 240 volt	nr		24.96		24.96
027	10 Litre, 3kW, 240 volt	nr		24.96		24.96
T3710	WATER HEATERS : supply only water heater, over or under sink application, swivel spout, thermostatically controlled					
025	7 Litre, 3kW, 240 volt	nr	179.31			179.31
027	10 Litre, 3kW, 240 volt	nr	221.26			221.26
T3715	WATER HEATERS : remove damaged water heater, supply and fix new water heater, over or under sink application, swivel spout, thermostatically controlled					
025	7 litre, 3kW, 240 volt	nr	179.31	24.96		204.27
027	10 litre, 3kW, 240 volt	nr	221.26	24.96		246.22
T3720	WATER HEATERS : fix only water heater, over or under sink application, swivel spout, thermostatically controlled					
035	7 Litre, 3 kW, 240 volt	nr		19.04		19.04
037	10 Litre, 3 kW, 240 volt	nr		19.04		19.04
T3725	WATER HEATERS : supply and fix water heater, over or under sink application, swivel spout, thermostatically controlled					
035	7 Litre, 3 kW, 240 volt	nr	179.31	19.04		198.35
037	10 Litre, 3 kW, 240 volt	nr	221.26	19.04		240.30
T3730	WATER HEATERS : fix only water heater, cistern type, electronic or immersion, thermostatically controlled					
040	25 Litre capacity	nr		25.38		25.38
041	50 Litre capacity	nr		38.07		38.07
042	75 Litre capacity	nr		38.07		38.07

© NSR 01 Aug 2008 - 31 Jul 2009   T1

T : ELECTRICAL SERVICES : WORKS OF ALTERATION/SMALL WORKS			Mat. £	Lab. £	Plant £	Total £
T3730						
043	3 kW, 125 Litre capacity	nr		50.76		50.76
044	6 kW, 125 Litre capacity	nr		50.76		50.76
T3735	WATER HEATERS : supply only water heater, electonic cistern type, thermostatically controlled					
040	25 Litre capacity	nr	809.00			809.00
041	50 Litre capacity	nr	959.00			959.00
042	75 Litre capacity	nr	1061.29			1061.29
043	3 kW, 125 Litre capacity	nr	1214.62			1214.62
044	6 kW, 125 Litre capacity	nr	1290.87			1290.87
T3740	WATER HEATERS : supply and fix water heater, cistern type, electronic or immersion, thermostatically controlled					
040	25 Litre capacity	nr	809.00	25.38		834.38
041	50 Litre capacity	nr	959.00	38.07		997.07
042	75 Litre capacity	nr	1061.29	38.07		1099.36
043	3 kW, 125 Litre capacity	nr	1214.62	50.76		1265.38
044	6 kW, 125 Litre capacity	nr	1290.87	50.76		1341.63
T3745	WATER HEATERS : supply only water heater, immersion cistern type, thermostatically controlled					
045	25 Litre capacity	nr	720.43			720.43
046	50 Litre capacity	nr	892.56			892.56
047	75 Litre capacity	nr	1061.29			1061.29
048	125 Litre capacity	nr	1214.62			1214.62
T3750	WATER HEATERS : supply and fix water heater, cistern type, electronic or immersion, thermostatically controlled					
045	25 Litre capacity	nr	720.43	25.38		745.81
046	50 Litre capacity	nr	892.56	38.07		930.63
047	75 Litre capacity	nr	1061.29	50.76		1112.05
048	125 Litre capacity	nr	1214.62	50.76		1265.38
T3755	ELECTRIC SHOWERS : fix only instantaneous electric shower, flexible hose, rail, shower head, or swivel spout					
050	7 - 10 kW Domestic type	nr		25.38		25.38
052	7 - 10 kW Industrial type	nr		31.72		31.72

© NSR 01 Aug 2008 - 31 Jul 2009   T2

T : ELECTRICAL SERVICES : WORKS OF ALTERATION/SMALL WORKS			Mat. £	Lab. £	Plant £	Total £
T3760	ELECTRIC SHOWERS : supply only instantaneous electric shower					
050	7 - 10 kW Domestic type	nr	256.11			256.11
052	7 - 10 kW Industrial type	nr	218.28			218.28
T3765	ELECTRIC SHOWERS : supply and fix instantaneous electric shower, flexible hose, rail, shower head, or swivel spout					
050	7 - 10 kW Domestic type	nr	256.11	25.38		281.49
052	7 - 10 kW Industrial type	nr	218.28	31.72		250.00
T3770	IMMERSION HEATERS : remove damaged item of equipment, supply and fix new item of equipment					
035	Immersion thermostat	nr	7.46	11.42		18.88
T3775	IMMERSION HEATERS : remove damaged immersion heater, fix only new immersion heater, final flexible connection					
050	1 - 4 kW, Single phase, 240 volt	nr		19.04		19.04
052	1 - 6 kW, Three phase, 415 volt	nr		25.38		25.38
T3780	IMMERSION HEATERS : supply only immersion heater					
050	1 - 4 kW, Single phase, 240 volt	nr	31.57			31.57
052	1 - 6 kW, Three phase, 415 volt	nr	160.00			160.00
T3785	IMMERSION HEATERS : remove damaged immersion heater, supply and fix new immersion heater, final flexible connection					
050	1 - 4 kW, single phase, 240 volt	nr	31.57	19.04		50.61
052	1 - 6 kW, three phase, 415 volt	nr	160.00	25.38		185.38
T3790	IMMERSION HEATERS : fix only immersion heater, final flexible connection					
070	1 - 4 kW, single phase	nr		12.69		12.69
072	1 - 6 kW, three phase	nr		16.92		16.92
T3795	IMMERSION HEATERS : supply and fix immersion heater, final flexible connection					
070	1 - 4 kW, single phase	nr	31.57	12.69		44.26
072	1 - 6 kW, three phase	nr	160.00	16.92		176.92

© NSR 01 Aug 2008 - 31 Jul 2009

T3

T : ELECTRICAL SERVICES : WORKS OF ALTERATION/SMALL WORKS		Mat. £	Lab. £	Plant £	Total £
© NSR 01 Aug 2008 - 31 Jul 2009					T4

			Mat. £	Lab. £	Plant £	Total £
	U : ELECTRICAL SERVICES : WORKS OF ALTERATION/SMALL WORKS					
	U40 : EXTRACT VENTILATION INSTALLATION					
U4000	**EXTRACT FANS : remove damaged extract fan, fix only new extract fan, final flexible connection**					
013	100 mm Diameter, window/wall mounting type	nr		26.65		26.65
021	150 mm Diameter, window/wall mounting type	nr		26.65		26.65
023	225 mm Diameter, window/wall mounting type	nr		29.61		29.61
025	300 mm Diameter, window/wall mounting type	nr		29.61		29.61
029	225 mm Diameter, roof mounting type	nr		44.41		44.41
031	300 mm Diameter, roof mounting type	nr		44.41		44.41
U4009	**EXTRACT FANS : supply only extract fan**					
013	100 mm Diameter, window/wall mounting type	nr	12.50			12.50
021	150 mm Diameter, window/wall mounting type	nr	102.50			102.50
023	225 mm Diameter, window/wall mounting type	nr	197.37			197.37
025	300 mm Diameter, window/wall mounting type	nr	302.36			302.36
029	225 mm Diameter, roof mounting type	nr	283.84			283.84
031	300 mm Diameter, roof mounting type	nr	374.69			374.69
U4018	**EXTRACT FANS : remove damaged extract fan, supply and fix new extract fan, final flexible connection**					
013	100 mm diameter, window/wall mounting type	nr	12.50	26.65		39.15
021	150 mm diameter, window/wall mounting type	nr	102.50	26.65		129.15
023	225 mm diameter, window/wall mounting type	nr	197.37	29.61		226.98
025	300 mm diameter, window/wall mounting type	nr	302.36	29.61		331.97
029	225 mm diameter, roof mounting type	nr	283.84	44.41		328.25
031	300 mm diameter, roof mounting type	nr	374.69	44.41		419.10
U4027	**EXTRACT FANS : fix only extract fan, final flexible connection**					
005	100 mm Diameter, window/wall mounting type	nr		21.15		21.15
014	150 mm Diameter, window/wall mounting type	nr		21.15		21.15
015	225 mm Diameter, window/wall mounting type	nr		31.72		31.72
017	300 mm Diameter, window/wall mounting type	nr		38.07		38.07
026	225 mm Diameter, roof mounting type	nr		38.92		38.92

© NSR 01 Aug 2008 - 31 Jul 2009   U1

U : ELECTRICAL SERVICES : WORKS OF ALTERATION/SMALL WORKS			Mat. £	Lab. £	Plant £	Total £
U4027						
027	300 mm Diameter, roof mounting type	nr		45.68		45.68
U4036	**EXTRACT FANS : supply and fix extract fan, final flexible connection**					
005	100 mm Diameter, window/wall mounting type	nr	12.50	21.15		33.65
014	150 mm Diameter, window/wall mounting type	nr	102.50	21.15		123.65
015	225 mm Diameter, window/wall mounting type	nr	197.37	31.72		229.09
017	300 mm Diameter, window/wall mounting type	nr	302.36	38.07		340.43
026	225 mm Diameter, roof mounting type	nr	283.84	38.92		322.76
027	300 mm Diameter, roof mounting type	nr	374.69	45.68		420.37
U4045	**EXTRACT FANS : clean out item**					
100	150 mm - 450 mm Diameter, all types	nr		8.46		8.46
U4054	**FAN CONTROLLERS : remove damaged fan controller, fix only new fan controller**					
735	On/off, single speed, reversing controller	nr		21.15		21.15
737	On/off, variable speed, reversing controller	nr		21.15		21.15
U4063	**FAN CONTROLLERS : supply only fan controller**					
735	On/off, single speed, reversing controller	nr	90.67			90.67
737	On/off, variable speed, reversing controller	nr	110.30			110.30
U4072	**FAN CONTROLLERS : remove damaged fan controller, supply and fix new fan controller**					
735	On/off, single speed, reversing fan controller	nr	90.67	21.15		111.82
737	On/off, variable speed, reversing fan controller	nr	110.30	21.15		131.45
U4081	**FAN CONTROLLERS : fix only fan controller**					
720	On/off, two speed reversing controller	nr		12.69		12.69
722	On/off, variable speed reversing controller	nr		12.69		12.69
U4090	**FAN CONTROLLERS : supply and fix fan controller**					
720	On/off, two speed reversing controller	nr	90.67	12.69		103.36
722	On/off, variable speed reversing controller	nr	110.30	12.69		122.99

© NSR 01 Aug 2008 - 31 Jul 2009 U2

			Mat. £	Lab. £	Plant £	Total £
	V : ELECTRICAL SERVICES : WORKS OF ALTERATION/SMALL WORKS					
	V10 : LV SWITCHGEAR AND DISTRIBUTION					
V1000	**REWIREABLE FUSES : remove ruptured fusewire, rewire fuse carrier**					
020	5 - 60 Amp range	nr		6.34		6.34
V1002	**HRC FUSES : remove ruptured fuse, supply and fix new HRC fuse**					
025	5 -32 Amp range	nr	6.45	6.34		12.79
027	32 - 50 Amp range	nr	12.95	6.34		19.29
029	62 - 80 Amp range	nr	13.70	6.34		20.04
031	80 - 100 Amp range	nr	23.75	6.34		30.09
V1004	**HRC FUSES : supply fuse, HRC, and fitting in fuseboard or item of equipment**					
010	5 - 6 Amp	nr	0.75	2.96		3.71
012	10 Amp	nr	0.99	2.96		3.95
014	15 Amp	nr	0.99	2.96		3.95
016	30 Amp	nr	1.11	2.96		4.07
018	40 Amp	nr	4.20	2.96		7.16
020	45 Amp	nr	4.20	2.96		7.16
022	50 Amp	nr	4.62	2.96		7.58
024	63 Amp	nr	6.68	2.96		9.64
026	80 Amp	nr	10.21	2.96		13.17
028	100 Amp	nr	10.31	2.96		13.27
V1006	**HRC FUSES : remove existing rewireable fuse and carrier from consumer unit or distribution board, install new HRC fuse and carrier**					
010	5 - 6 Amp	nr	6.98	2.96		9.94
012	10 Amp	nr	6.98	2.96		9.94
014	16 Amp	nr	6.30	2.96		9.26
016	32 Amp	nr	9.19	2.96		12.15
018	40 Amp	nr	16.49	2.96		19.45
020	45 Amp	nr	16.49	2.96		19.45
022	50 Amp	nr	16.49	2.96		19.45

© NSR 01 Aug 2008 - 31 Jul 2009

V1

V : ELECTRICAL SERVICES : WORKS OF ALTERATION/SMALL WORKS			Mat. £	Lab. £	Plant £	Total £
V1006						
024	60 Amp	nr	16.49	2.96		19.45
026	80 Amp	nr	37.01	2.96		39.97
028	100 Amp	nr	37.01	2.96		39.97
V1008	**MCB'S : remove damaged MCB, supply and fix new MCB**					
035	SP, 5 - 32 Amp	nr	10.21	7.61		17.82
037	SP, 32 - 62 Amp	nr	10.77	7.61		18.38
039	TP, 5 - 32 Amp	nr	41.76	8.46		50.22
041	TP, 32 - 62 Amp	nr	45.42	8.46		53.88
043	TP, 80 - 100 Amp	nr	83.78	8.88		92.66
V1010	**MCB'S : miniature circuit breaker, fitting in distribution equipment, connections**					
060	SP, 6-32 amp	nr	10.21	2.96		13.17
062	SP, 40 amp	nr	10.77	2.96		13.73
064	SP, 45 amp	nr	10.77	2.96		13.73
066	SP, 50 amp	nr	10.77	2.96		13.73
068	SP, 63 amp	nr	10.48	2.96		13.44
070	TP, 6 amp	nr	44.14	2.96		47.10
072	SPN/DP, 20 Amp	nr	27.63	2.96		30.59
074	SPN/DP, 40 Amp	nr	29.19	2.96		32.15
076	SPN/DP, 63 Amp	nr	29.19	2.96		32.15
078	TP/TPN, 16 Amp	nr	42.97	2.96		45.93
080	TP/TPN, 50 Amp	nr	44.14	2.96		47.10
082	TP/TPN, 63 Amp	nr	29.19	2.96		32.15
084	TP, 30 amp	nr	75.88	2.96		78.84
086	TP, 40 amp	nr	79.47	2.96		82.43
088	TP, 50 amp	nr	79.47	2.96		82.43
090	TP, 63 amp	nr	44.14	2.96		47.10
092	TP, 80 amp	nr	81.16	2.96		84.12
094	TP, 100 amp	nr	83.78	2.96		86.74
V1012	**MCCB'S : remove damaged MCCB, supply and fix new MCCB**					
050	DP, 15 - 100 Amp	nr	65.00	8.88		73.88

© NSR 01 Aug 2008 - 31 Jul 2009

V2

V : ELECTRICAL SERVICES : WORKS OF ALTERATION/SMALL WORKS			Mat. £	Lab. £	Plant £	Total £
V1012						
056	TPN, 15 - 100 Amp	nr	140.00	8.88		148.88
V1014	**MCCB'S : moulded case circuit breaker, fitting in distribution equipment, connections**					
100	SP, 15 - 30 amp	nr	60.00	3.81		63.81
102	SP, 40 - 60 amp	nr	65.00	3.81		68.81
104	SP, 70 - 100 amp	nr	65.00	3.81		68.81
106	TPN, 15 - 32 amp	nr	130.00	3.81		133.81
108	TPN, 40 - 60 amp	nr	130.00	3.81		133.81
110	TPN, 70 - 100 amp	nr	140.00	3.81		143.81
V1016	**MCB/RCD'S : miniature circuit breaker/residual current operated circuit breaker, fitting in distribution equipment, connections**					
115	SP, 6 amp, 30 mA sensitivity	nr	68.46	2.11		70.57
117	SP, 20 amp, 30 mA sensitivity	nr	68.46	2.11		70.57
119	SP, 32 amp, 30 mA sensitivity	nr	68.46	2.11		70.57
121	SP, 40 amp, 30 mA sensitivity	nr	68.46	2.11		70.57
123	DP, 20amp, 30 mA sensitivity	nr	71.33	2.11		73.44
125	DP, 40 amp, 30 mA sensitivity	nr	73.43	2.11		75.54
127	DP, 63 amp, 30 mA sensitivity	nr	62.76	2.11		64.87
129	DP, 100 amp, 30 mA sensitivity	nr	101.96	2.11		104.07
131	TP, 6 amp, 30 mA sensitivity	nr	70.52	2.11		72.63
133	TP, 10 amp, 30 mA sensitivity	nr	70.52	2.11		72.63
135	TP, 15 amp, 30 mA sensitivity	nr	70.52	2.11		72.63
137	TP, 20 amp, 30 mA sensitivity	nr	70.52	2.11		72.63
139	TP, 32 amp, 30 mA sensitivity	nr	70.52	2.11		72.63
141	TP, 40 amp, 30 mA sensitivity	nr	70.52	2.11		72.63
143	TP, 50 amp, 30 mA sensitivity	nr	91.42	2.11		93.53
145	TP, 63 amp, 30 mA sensitivity	nr	90.78	2.11		92.89
V1018	**RCD'S : remove damaged RCD, supply and fix new RCD**					
070	DP, 5 - 30 Amp, 10 - 30 ma	nr	105.30	9.73		115.03
V1020	**RCD'S : residual current device, fitting in distribution equipment, connections**					
150	2 Pole, 16 - 32 amp	nr	58.70	2.96		61.66

© NSR 01 Aug 2008 - 31 Jul 2009   V3

	V : ELECTRICAL SERVICES : WORKS OF ALTERATION/SMALL WORKS			Mat. £	Lab. £	Plant £	Total £
V1020							
152	2 Pole, 35 - 63 amp		nr	63.88	4.23		68.11
154	2 Pole, 80 - 100 amp		nr	83.39	4.23		87.62
156	4 Pole, 16 - 32 amp		nr	74.07	5.08		79.15
158	4 Pole, 35 - 63 amp		nr	82.46	5.08		87.54
160	4 Pole, 80 - 100 amp		nr	135.64	5.08		140.72
V1022	RCD'S : fix only residual current operated circuit breaker complete with enclosure, designation label and phase discs						
162	2/4 Pole, 16 - 100 amp		nr		19.04		19.04
V1024	RCD'S : supply only RCD						
150	2 Pole, 16 - 32 amp		nr	65.92			65.92
152	2 Pole, 35 - 63 amp		nr	71.10			71.10
154	2 Pole, 80 - 100 amp		nr	90.61			90.61
156	4 Pole, 16 - 32 amp		nr	85.08			85.08
158	4 Pole, 35 - 63 amp		nr	93.47			93.47
160	4 Pole, 80 - 100 amp		nr	146.65			146.65
V1026	RCD'S : supply and fix residual current operated circuit breaker complete with enclosure, designation label and phase discs						
150	2 Pole, 16 - 32 amp		nr	65.92	19.04		84.96
152	2 Pole, 35 - 63 amp		nr	71.10	19.04		90.14
154	2 Pole, 80 - 100 amp		nr	90.61	19.04		109.65
156	4 Pole, 16 - 32 amp		nr	85.08	19.04		104.12
158	4 Pole, 35 - 63 amp		nr	93.47	19.04		112.51
160	4 Pole, 80 - 100 amp		nr	146.65	19.04		165.69
V1028	ISOLATORS : remove damaged isolator, fix only new isolator, designation label and phase discs						
072	SPN/DP, 10 - 32 Amp		nr		25.38		25.38
074	SPN/DP, 35 - 63 Amp		nr		35.53		35.53
076	SPN/DP, 80 - 100 Amp		nr		81.95		81.95
078	TP/TPN, 10 - 32 Amp		nr		40.61		40.61
080	TP/TPN, 35 - 63 Amp		nr		91.06		91.06
082	TP/TPN, 80 - 100 Amp		nr		91.06		91.06

© NSR 01 Aug 2008 - 31 Jul 2009

V : ELECTRICAL SERVICES : WORKS OF ALTERATION/SMALL WORKS			Mat. £	Lab. £	Plant £	Total £
V1030	ISOLATORS : fix only isolator, designation label and phase discs					
180	SPN/DP, 10 - 32 amp	nr		12.69		12.69
182	SPN/DP, 35 - 63 amp	nr		17.77		17.77
184	SPN/DP, 80 - 100 amp	nr		22.84		22.84
186	TP/TPN, 10 - 32 amp	nr		20.30		20.30
188	TP/TPN, 35 - 63 amp	nr		25.38		25.38
190	TP/TPN, 80 - 100 amp	nr		38.07		38.07
V1032	ISOLATORS : supply and fix isolator, designation label and phase discs					
180	SPN/DP, 10 - 32 amp	nr	67.82	12.69		80.51
182	SPN/DP, 35 - 63 amp	nr	117.39	17.77		135.16
184	SPN/DP, 80 -100 amp	nr	178.50	22.84		201.34
186	TP/TPN, 10 - 32 amp	nr	77.88	20.30		98.18
188	TP/TPN, 35 - 63 amp	nr	149.46	25.38		174.84
190	TP/TPN, 80 - 100 amp	nr	235.75	38.07		273.82
V1034	ISOLATORS : supply only new isolator, designation label and phase discs					
072	SPN/DP, 10 - 32 Amp	nr	67.82			67.82
074	SPN/DP, 35 - 63 Amp	nr	117.39			117.39
076	SPN/DP, 80 - 100 Amp	nr	178.50			178.50
078	TP/TPN, 10 - 32 Amp	nr	77.88			77.88
080	TP/TPN, 35 - 63 Amp	nr	149.46			149.46
082	TP/TPN, 80 - 100 Amp	nr	235.75			235.75
V1036	ISOLATORS : remove damaged isolator, supply and fix new isolator, designation label and phase discs.					
072	SPN/DP, 10 - 32 Amp	nr	67.82	25.38		93.20
074	SPN/DP, 35 - 63 Amp	nr	117.39	35.53		152.92
076	SPN/DP, 80 - 100 Amp	nr	178.50	81.95		260.45
078	TP/TPN, 10 - 32 Amp	nr	77.88	40.61		118.49
080	TP/TPN, 35 - 63 Amp	nr	149.46	91.06		240.52
082	TP/TPN, 80 - 100 Amp	nr	235.75	76.14		311.89

© NSR 01 Aug 2008 - 31 Jul 2009

V5

V : ELECTRICAL SERVICES : WORKS OF ALTERATION/SMALL WORKS			Mat. £	Lab. £	Plant £	Total £
V1038	FUSED SWITCHES : remove damaged fused switch, fix only new fused switch, HRC fuse, designation label and phase discs					
074	SPN/DP, 35 - 63 Amp	nr		50.76		50.76
076	SPN/DP, 80 - 100 Amp	nr		118.38		118.38
080	TP/TPN, 35 - 63 Amp	nr		55.84		55.84
082	TP/TPN, 80 - 100 Amp	nr		136.59		136.59
V1040	FUSED SWITCHES : fix only fused switch, HRC fuse, designation label and phase discs					
182	SPN/DP, 35 - 63 amp	nr		25.38		25.38
184	SPN/DP, 80 - 100 amp	nr		59.19		59.19
188	TP/TPN, 35 - 63 amp	nr		27.92		27.92
190	TP/TPN, 80 - 100 amp	nr		68.29		68.29
V1042	FUSED SWITCHES : supply only fused switch					
074	SPN/DP, 35 - 63 Amp	nr	180.96			180.96
076	SPN/DP, 80 - 100 Amp	nr	309.39			309.39
080	TP/TPN, 35 - 63 Amp	nr	232.40			232.40
082	TP/TPN, 80 - 100 Amp	nr	407.09			407.09
V1044	FUSED SWITCHES : remove damaged fused switch, supply and fix new fused switch, HRC fuse, designation label and phase discs					
072	SPN/DP, 10 - 32 Amp	nr	87.69	35.53		123.22
074	SPN/DP, 35 - 63 Amp	nr	180.96	50.76		231.72
076	SPN/DP, 80 - 100 Amp	nr	309.39	118.38		427.77
080	TP/TPN, 35 - 63 Amp	nr	232.40	55.84		288.24
082	TP/TPN, 80 - 100 Amp	nr	407.09	136.59		543.68
V1046	FUSED SWITCHES : supply and fix fused switch, HRC fuse, designation label and phase discs					
082	SPN/DP, 35 - 63 amp	nr	180.96	25.38		206.34
084	SPN/DP/, 80 - 100 amp	nr	309.39	59.19		368.58
088	TP/TPN, 35 - 63 amp	nr	232.40	27.92		260.32
090	TP/TPN, 80 - 100 amp	nr	407.09	68.29		475.38
V1048	SWITCHED FUSES : removed damaged switched fuse, fix only new switched fuse, HRC fuse, designation label and phase discs					
072	SPN/DP, 10 - 32 Amp	nr		35.53		35.53

© NSR 01 Aug 2008 - 31 Jul 2009

V6

V : ELECTRICAL SERVICES : WORKS OF ALTERATION/SMALL WORKS			Mat. £	Lab. £	Plant £	Total £
V1048						
074	SPN/DP, 35 - 63 Amp	nr		50.76		50.76
076	SPN/DP, 80 - 100 Amp	nr		118.38		118.38
078	TP/TPN, 10 - 32 Amp	nr		55.84		55.84
080	TP/TPN, 35 - 63 Amp	nr		100.17		100.17
082	TP/TPN, 80 - 100 Amp	nr		136.59		136.59
V1050	SWITCHED FUSES : fix only switched fuse, HRC fuse, designation label and phase discs					
072	SPN/DP, 10 - 32 Amp	nr		17.77		17.77
074	SPN/DP, 35 - 63 Amp	nr		25.38		25.38
076	SPN/DP, 80 - 100 Amp	nr		59.19		59.19
078	TP/TPN, 10 - 32 Amp	nr		27.92		27.92
080	TP/TPN, 35 - 63 Amp	nr		50.08		50.08
082	TP/TPN, 80 - 100 Amp	nr		68.29		68.29
V1052	SWITCHED FUSES : supply only switched fuse					
072	SPN/DP, 10 - 32 Amp	nr	87.69			87.69
074	SPN/DP, 35 - 63 Amp	nr	180.96			180.96
076	SPN/DP, 80 - 100 Amp	nr	309.39			309.39
078	TP/TPN, 10 - 32 Amp	nr	106.15			106.15
080	TP/TPN, 35 - 63 Amp	nr	232.40			232.40
082	TP/TPN, 80 - 100 Amp	nr	407.09			407.09
V1054	SWITCHED FUSES : removed damaged switched fuse, supply and fix new switched fuse, HRC fuse, designation label and phase discs					
072	SPN/DP, 10 - 32 Amp	nr	87.69	35.53		123.22
074	SPN/DP, 35 - 63 Amp	nr	180.96	50.76		231.72
076	SPN/DP, 80 - 100 Amp	nr	309.39	118.38		427.77
078	TP/TPN, 10 - 32 Amp	nr	106.15	55.84		161.99
080	TP/TPN, 35 - 63 Amp	nr	232.40	100.17		332.57
082	TP/TPN, 80 - 100 Amp	nr	407.09	136.59		543.68
V1056	SWITCHED FUSES : supply and fix switched fuse, HRC fuse, designation label and phase discs					
072	SPN/DP, 10 - 32 amp	nr	87.69	17.77		105.46

© NSR 01 Aug 2008 - 31 Jul 2009

V : ELECTRICAL SERVICES : WORKS OF ALTERATION/SMALL WORKS			Mat. £	Lab. £	Plant £	Total £
V1056						
074	SPN/DP, 35 - 63 amp	nr	180.96	25.38		206.34
076	SPN/DP, 80 - 100 amp	nr	309.39	59.19		368.58
078	TP/TPN, 10 - 32 Amp	nr	106.15	27.92		134.07
080	TP/TPN, 35 - 63 Amp	nr	232.40	50.08		282.48
082	TP/TPN, 80 - 100 Amp	nr	407.09	38.07		445.16
V1058	CONSUMER UNITS : remove damaged distribution board or consumer unit with HRC fuseways, or MCB's, or RCD's or RCCB's, fix only new distribution board or consumer unit with incoming devices, designation labels, phase discs and circuit charts (outgoing protection devices and sub circuit testing measured separately)					
100	One - three way, SPN/DP	nr		96.44		96.44
102	Four - six way, SPN/DP	nr		107.44		107.44
104	Eight - twelve way, SPN/DP	nr		118.44		118.44
106	Two - four way, TPN	nr		118.44		118.44
108	Six - eight way, TPN	nr		198.81		198.81
110	Ten - twelve way, TPN	nr		287.64		287.64
V1060	CONSUMER UNITS : fix only distribution boards or consumer unit with incoming protection device, designation labels, phase discs and circuit charts (outgoing protection devices and sub circuit testing measured separately)					
210	One - three way, SPN/DP	nr		48.22		48.22
212	Four - six way, SPN/DP	nr		53.72		53.72
214	Eight - twelve way, SPN/DP	nr		59.22		59.22
216	Two - four way, TPN	nr		59.22		59.22
218	Six - eight way, TPN	nr		99.41		99.41
220	Ten - twelve way, TPN	nr		139.59		139.59
V1062	CONSUMER UNITS : supply only new distribution board or consumer unit with incoming isolator switch, designation labels, phase discs and circuit charts (outgoing protection devices and sub circuit testing measured separately)					
100	One - three way, SPN/DP	nr	14.95			14.95
102	Four - six way, SPN/DP	nr	17.95			17.95
104	Eight - twelve way, SPN/DP	nr	19.95			19.95
106	Two - four way, TPN	nr	133.19			133.19
108	Six - eight way, TPN	nr	154.46			154.46

© NSR 01 Aug 2008 - 31 Jul 2009

V8

	V : ELECTRICAL SERVICES : WORKS OF ALTERATION/SMALL WORKS		Mat. £	Lab. £	Plant £	Total £
V1062						
110	Ten - twelve way, TPN	nr	198.66			198.66
V1064	CONSUMER UNITS : remove damaged distribution board or consumer unit with HRC fuseways, or MCB's, or RCD's or RCCB's, supply and fix new distribution board or consumer unit with incoming devices, designation labels, phase discs and circuit charts (outgoing protection devices and sub circuit testing measured separately)					
100	One - three way, SPN/DP	nr	14.95	96.44		111.39
102	Four - six way, SPN/DP	nr	17.95	107.44		125.39
104	Eight - twelve way, SPN/DP	nr	19.95	118.44		138.39
106	Two - four way, TPN	nr	133.19	118.44		251.63
108	Six - eight way, TPN	nr	173.40	198.81		372.21
11O	Ten - twelve way, TPN	nr	198.66	279.18		477.84
V1066	CONSUMER UNITS : supply and fix distribution board or consumer unit with incoming protection device, designation labels, phase discs and circuit charts (outgoing protection devices and sub circuit testing measured separately)					
110	One - three way, SPN/DP	nr	14.95	48.22		63.17
112	Four - six way, SPN/DP	nr	17.95	53.72		71.67
114	Eight - twelve way, SPN/DP	nr	19.95	59.22		79.17
116	Two - four way, TPN	nr	133.19	59.22		192.41
118	Six - eight way, TPN	nr	154.46	99.41		253.87
120	Ten - twelve way, TPN	nr	198.66	139.59		338.25
V1068	CONSUMER UNITS: supply only new distribution board or consumer unit with 30mA or 100mA incoming RCD device, designation labels, phase discs and circuit charts (outgoing protection devices and sub circuit testing measured separately)					
090	Four way SPN/DP distribution board/ consumer unit 100A.	nr	72.63			72.63
101	Six way SPN/DP distribution board/consumer unit 100A	nr	78.93			78.93
105	Eight way SPN/DP distribution board/ consumer unit 100A	nr	82.50			82.50
109	Twelve way SPN/DP distribution board/ consumer unit 100A	nr	98.99			98.99
112	Two - four way TPN distribution board/ consumer unit 100mA RCD	nr	406.31			406.31
114	Six - eight way TPN distribution board/ consumer unit 100mA RCD	nr	442.14			442.14
115	Ten - twelve way TPN distribution board/ consumer unit 100mA RCD	nr	480.79			480.79

© NSR 01 Aug 2008 - 31 Jul 2009   V9

V : ELECTRICAL SERVICES : WORKS OF ALTERATION/SMALL WORKS			Mat. £	Lab. £	Plant £	Total £
V1070	CONSUMER UNITS : remove damaged distribution board, supply and fix new distribution board or consumer unit with 30mA or 100mA incoming RCD device, designation labels, phase discs and circuit charts (outgoing protection devices and sub circuit testing measured separately)					
090	Four way, SPN/DP distribution board or consumer unit, 100A	nr	72.63	96.44		169.07
101	Six way, SPN/DP distribution board or consumer unit 100A	nr	78.93	107.44		186.37
105	Eight way, SPN/DP distribution board or consumer unit, 100A	nr	82.50	118.44		200.94
109	Twelve way, SPN/DP distribution board or consumer unit, 100A	nr	98.99	118.44		217.43
112	Two - four way TPN distribution board or consumer unit, 100A	nr	442.14	118.44		560.58
114	Six - eight way TPN distribution board or consumer unit, 100A	nr	442.14	198.81		640.95
115	Ten - twelve way TPN distribution board or consumer unit, 100A	nr	480.79	279.18		759.97
V1072	CONSUMER UNITS: supply and fix new distribution board or consumer unit with 30mA or 100mA incoming RCD device, designation labels, phase discs and circuit charts (outgoing protection devices and sub circuit testing measured separately)					
108	Three way SPN/DP distribution board/consumer unit 40A	nr	72.63	48.22		120.85
109	Three way SPN/DP distribution board/consumer unit 80A	nr	72.63	48.22		120.85
111	Five way SPN/DP distribution board/consumer unit 40A	nr	78.93	53.72		132.65
113	Five way SPN/DP distribution board/consumer unit 80A	nr	78.93	53.72		132.65
117	Seven way SPN/DP distribution board/consumer unit 40A	nr	82.50	59.22		141.72
119	Seven way SPN/DP distribution board/consumer unit 80A	nr	82.50	59.22		141.72
121	Eleven way SPN/DP distribution board/consumer unit 80A	nr	98.99	59.22		158.21
123	Two - four way TPN distribution board/consumer unit 100mA	nr	406.31	59.22		465.53
124	Six - eight way TPN distribution board/consumer unit 100mA	nr	442.14	99.41		541.55
125	Ten - twelve way TPN distribution board/consumer unit 100mA	nr	480.79	139.59		620.38
V1074	SPLIT LOAD CONSUMER UNITS : remove damaged split load distribution board or consumer unit with HRC fuseways, or MCB's, or RCD's, or RCCB's, fix only new split load distribution board or consumer unit with incoming protection devices, designation labels, phase discs and circuit charts (outgoing protection devices and sub circuit testing measured separately)					
116	Five MCB ways, 2/3 RCD ways, 30 mAmp	nr		107.44		107.44
118	Five MCB ways, 2/3 RCD ways, 100 mAmp	nr		107.44		107.44
120	Nine MCB ways, 3 - 6 RCD ways, 30 mAmp	nr		161.16		161.16
122	Nine MCB ways, 3 - 6 RCD ways, 100 mAmp	nr		161.16		161.16

© NSR 01 Aug 2008 - 31 Jul 2009

V10

	V : ELECTRICAL SERVICES : WORKS OF ALTERATION/SMALL WORKS		Mat. £	Lab. £	Plant £	Total £
V1076	SPLIT LOAD CONSUMER UNITS : supply only new split load distribution board or consumer unit with incoming RCD protection devices, designation labels, phase discs and circuit charts (outgoing protection devices and sub circuit testing measured separately)					
116	Nine MCB ways, 4 - 5 RCD ways, 100mAmp 30 mAmp	nr	185.78			185.78
118	Twelve MCB ways, 7 - 5 RCD ways, 100mAmp/30 mAmp	nr	193.79			193.79
120	Twelve MCB ways, 6 - 6 RCD ways, 100mAmp -30 mAmp	nr	193.79			193.79
V1078	SPLIT LOAD CONSUMER UNITS : remove damaged split load distribution board or consumer unit with HrC fuseways, or MCB's, or RCD's, or RCCB's, supply and fix new split load distribution board or consumer unit with incoming protection devices, designation labels, phase discs and circuit charts (outgoing protection devices and sub circuit testing measured separately)					
116	Nine MCB ways, 5 - 4 RCD ways, 100mAmp - 30 mAmp	nr	185.78	161.16		346.94
120	Twelve MCB ways, 7 - 5 RCD ways, 100mAmp - 30 mAmp	nr	193.79	161.16		354.95
122	Twelve MCB ways, 6 - 6 RCD ways, 100 mAmp - 30mAmp	nr	193.79	161.16		354.95
V1080	SPLIT LOAD CONSUMER UNITS : fix only split load distribution board or consumer unit with incoming protection devices, designation labels, phase discs and circuit charts (outgoing protection devices and sub circuit testing measured separately)					
230	Five MCB ways, 2/3 RCD ways, 30 mA	nr		53.72		53.72
232	Nine MCB ways, 5/4 RCD ways, 100 mA	nr		53.72		53.72
234	Twelve MCB ways, 7 - 5 RCD ways, 100 mA	nr		80.37		80.37
236	Twelve MCB ways, 6 - 6 RCD ways, 100 mA	nr		80.37		80.37
V1082	SPLIT LOAD CONSUMER UNITS : supply and fix split load distribution board or consumer unit with incoming protection devices, designation labels, phase discs and circuit charts (outgoing protection devices and sub circuit testing measured separately)					
232	Nine MCB ways, 5/4 RCD ways, 100 mA	nr	185.78	53.72		239.50
234	Twelve MCB ways, 7 - 5 RCD ways, 100 mA	nr	193.79	80.37		274.16
236	Twelve MCB ways, 6 - 6 RCD ways, 100 mA	nr	193.79	80.37		274.16
	V15 : LIGHTING INSTALLATION : LUMINAIRES					
V1500	TUNGSTEN LUMINAIRES : batten lampholder, moulded plastic					
005	3 Plate type, BC	nr	2.40	5.08		7.48

© NSR 01 Aug 2008 - 31 Jul 2009 — V11

V : ELECTRICAL SERVICES : WORKS OF ALTERATION/SMALL WORKS				Mat. £	Lab. £	Plant £	Total £
V1503	TUNGSTEN LUMINAIRES : remove damaged item of equipment, supply and fix new item of equipment						
016	Lampholders, pendant set		nr	3.67	19.46		23.13
018	Lampholders, batton type		nr	5.07	19.46		24.53
020	Lampholders, bayonet cap		nr	2.18	15.23		17.41
022	Lampholders, edison screw		nr	0.73	15.23		15.96
023	Lamps, 50 watt, LV/DICHROIC		nr	3.95	2.11		6.06
024	Lamps, 25 - 200 watt, GLS, BC/ES		nr	0.57	2.11		2.68
025	Lamps, 100 watt, halogen		nr	1.35	2.11		3.46
026	Lamps, 150 watt, halogen		nr	1.35	2.11		3.46
V1506	TUNGSTEN LUMINAIRES : pendant fitting comprising ceiling rose, flexible cable and lampholder						
005	3 Plate type, BC		nr	6.29	8.46		14.75
V1509	TUNGSTEN LUMINAIRES : remove damaged luminaire, fix only new luminaire, lamp, diffuser or the like, connectors and final flexible connection						
027	Lamps, 100 - 1000 watt, halogen		nr		36.80		36.80
030	Bulkhead fitting, 25 - 200 watt lamp		nr		36.80		36.80
032	Bulkhead fitting, 2 x 25 - 200 watt lamps		nr		36.80		36.80
034	Downlight fitting, 25 - 200 watt lamp		nr		36.80		36.80
036	Semi-recessed downlight, 25 - 200 watt lamp		nr		36.80		36.80
038	Recessed downlight, 25 - 200 watt lamp		nr		36.80		36.80
040	Pendant fitting, 25 - 200 watt lamp		nr		19.46		19.46
V1512	TUNGSTEN LUMINAIRES : supply only luminaire and lamps						
030	Bulkhead fitting, 25 - 200 watt lamp		nr	68.30			68.30
032	Bulkhead fitting, 2 x 25 - 200 watt lamps		nr	97.31			97.31
034	Downlight fitting, 25 - 200 watt lamp		nr	4.09			4.09
036	Semi-recessed downlight, 25 - 200 watt lamp		nr	30.00			30.00
038	Recessed downlight, 25 - 200 watt lamp		nr	26.50			26.50
040	Pendant fitting, 25 - 200 watt lamp		nr	6.59			6.59
044	Bulkhead fitting, IP65 16 watt lamp		nr	20.75			20.75
046	Bulkhead fitting, IP65 28 watt lamp		nr	31.10			31.10

© NSR 01 Aug 2008 - 31 Jul 2009   V12

V : ELECTRICAL SERVICES : WORKS OF ALTERATION/SMALL WORKS			Mat. £	Lab. £	Plant £	Total £
V1515	TUNGSTEN LUMINAIRES : fix only luminaire, lamp, diffuser or the like, connectors and final flexible connection					
020	Bulkhead fitting, 25 - 200 watt lamp	nr		30.46		30.46
022	Bulkhead fitting, 2 x 25 - 200 watt lamps	nr		30.46		30.46
024	Downlight fitting, 25 - 200 watt lamp	nr		30.46		30.46
026	Recessed downlight, 25 - 200 watt lamp	nr		30.46		30.46
028	Pendant fitting, 25 - 200 watt lamp	nr		15.23		15.23
044	Bulkhead fitting, IP65 16 watt lamp	nr		27.49		27.49
046	Bulkhead fitting, IP65 28 watt lamp	nr		27.49		27.49
V1518	TUNGSTEN LUMINAIRES : remove damaged luminaire, supply and fix new luminaire, lamp, diffuser or the like, connectors and final flexible connection					
030	Bulkhead fitting, 25 - 200 watt	nr	68.30	36.80		105.10
032	Bulkhead fitting, 2 x 25 - 200 watt	nr	97.31	36.80		134.11
034	Downlight fitting, 25 - 200 watt	nr	4.09	36.80		40.89
036	Semi-recessed downlight, 25 - 200 watt	nr	30.00	36.80		66.80
038	Recessed downlight, 25 - 200 watt	nr	26.50	36.80		63.30
040	Pendant fitting, 25 - 200 watt	nr	6.59	19.46		26.05
044	Bulkhead fitting, IP65 16 watt lamp	nr	20.75	33.84		54.59
046	Bulkhead fitting, IP65 28 watt lamp	nr	31.10	33.84		64.94
V1521	TUNGSTEN LUMINAIRES : supply and fix luminaire, lamp, diffuser or the like, connectors and final flexible connection					
020	Bulkhead fitting, 25 - 200 watt lamp	nr	68.30	30.46		98.76
022	Bulkhead fitting, 2 x 25 - 200 watt lamps	nr	97.31	30.46		127.77
024	Downlight fitting, 25 - 200 watt lamp	nr	4.09	30.46		34.55
026	Recessed downlight, 25 - 200 watt lamp	nr	26.50	30.46		56.96
028	Pendant fitting, 25 - 200 watt lamp	nr	6.59	15.23		21.82
044	Bulkhead fitting, IP65 16 watt lamp	nr	20.75	27.49		48.24
046	Bulkhead fitting, IP65 28 watt lamp	nr	31.10	27.49		58.59
V1524	FLUORESCENT LUMINAIRES : remove damaged item of equipment, supply and fix new item of equipment. (Removal and refitting of diffusers measured seperately)					
050	Lampholder, bi-pin, 16 mm diameter	nr	3.13	4.23		7.36
052	Lampholder, bi-pin, 26 mm diameter	nr	3.13	4.23		7.36
054	Lampholder, bi-pin, 38 mm diameter	nr	3.13	4.23		7.36

© NSR 01 Aug 2008 - 31 Jul 2009

V13

V : ELECTRICAL SERVICES : WORKS OF ALTERATION/SMALL WORKS			Mat. £	Lab. £	Plant £	Total £
V1524						
056	Lampholder, bi-pin, U tube	nr	3.13	5.08		8.21
058	Lampholder, circular tube	nr	3.13	5.08		8.21
059	Lamp, 100w halogen	nr	1.35	2.11		3.46
060	Lamp, 150w halogen	nr	1.35	2.11		3.46
062	Lamp, 300 mm, 8 watt, bi-pin	nr	1.35	2.11		3.46
064	Lamp, 600 mm, 18 watt, bi-pin	nr	2.05	2.11		4.16
066	Lamp, 900 mm, 30 watt, bi-pin	nr	3.05	2.11		5.16
068	Lamp, 1200 mm, 36 - 40 watt, bi-pin	nr	2.10	2.11		4.21
070	Lamp, 1500 mm, 58 - 80 watt, bi-pin	nr	2.35	2.11		4.46
072	Lamp, 1800 mm, 70 - 85 watt, bi-pin	nr	3.95	2.11		6.06
074	Lamp, 2400 mm, 100 - 125 watt, bi-pin	nr	6.25	2.11		8.36
076	Lamp, U tube, 40 watt, bi-pin	nr	7.28	2.11		9.39
078	Lamp, compact 2D, 16 - 38 watt	nr	6.75	2.11		8.86
079	Lamp, PL, 9 - 18 watt	nr	7.88	2.11		9.99
080	Lamp, SL, 10 - 26 watt	nr	5.87	2.11		7.98
082	Starter switch, 15 - 125 watt	nr	4.63	2.11		6.74
084	Capacitor, 4 - 8 microfarads, 250 volt	nr	4.30	8.46		12.76
086	Choke, 18 - 20 watt, 240 volt, one lamp	nr	6.65	10.15		16.80
088	Choke, 18 - 20 watt, 240 volt, two lamp	nr	13.30	10.15		23.45
090	Choke, 28 watt, 240 volt, one lamp	nr	7.20	10.15		17.35
092	Choke, 30 watt, 240 volt, one lamp	nr	7.20	10.15		17.35
094	Choke, 30 watt, 240 volt, two lamp	nr	14.40	10.15		24.55
096	Choke, 100 watt, 240 volt, one lamp	nr	9.95	10.15		20.10
098	Choke, 100 watt, 240 volt, two lamp	nr	19.90	10.15		30.05
100	Choke, 38 watt, 240 volt, one lamp	nr	7.20	10.15		17.35
105	Electronic ballast unit, 36 - 75 watt, 240 volt, one lamp	nr	60.40	10.15		70.55
107	Electronic ballast unit, 36 - 40 watt, 240 volt, two lamp	nr	60.40	10.15		70.55
120	Transtar high power unit	nr	36.73	2.11		38.84
V1527	FLUORESCENT LUMINAIRES : remove damaged luminaire, fix only new luminaire, lamp, diffuser, reflector or the like, connectors and final flexible connection					
160	Bulkhead fitting, 300 mm, 8 watt lamp	nr		36.80		36.80
162	Bulkhead fitting, 300 mm, 2 x 8 watt lamp	nr		36.80		36.80

© NSR 01 Aug 2008 - 31 Jul 2009 — V14

	V : ELECTRICAL SERVICES : WORKS OF ALTERATION/SMALL WORKS		Mat. £	Lab. £	Plant £	Total £
V1527						
164	Bulkhead fitting, 16 x 38 watt 2D lamp	nr		36.80		36.80
170	Batten fitting, 1500 mm, single, 58 watt lamp	nr		42.30		42.30
172	Batten fitting, 1800 mm, single, 70 watt lamp	nr		42.30		42.30
180	Recessed fitting, 1200 - 2400 mm, 36 - 40 watt lamp	nr		69.80		69.80
190	Circular fitting, 22 watt lamp	nr		42.30		42.30
192	Circular fitting, 40 watt lamp	nr		42.30		42.30
V1530	FLUORESCENT LUMINAIRES : supply only item, including diffuser					
160	Bulkhead fitting, 2 x 8 watt lamp	nr	52.67			52.67
164	Bulkhead fitting, 38 watt 2D lamp	nr	35.20			35.20
168	Batten fitting, 1200 mm, single, 36 watt lamp	nr	45.57			45.57
169	Batten fitting, 1200 mm, twin, 36 watt lamp	nr	78.11			78.11
170	Batten fitting, 1500 mm, single, 58 watt lamp	nr	47.26			47.26
171	Batten fitting, 1500 mm, twin, 58 watt lamp	nr	83.03			83.03
172	Batten fitting, 1800 mm, single, 70 watt lamp	nr	57.72			57.72
173	Batten fitting, 1800 mm, twin, 70 watt lamp	nr	93.49			93.49
174	Batten fitting, 2400 mm, single, 100 watt lamp	nr	112.77			112.77
175	Batten fitting, 2400 mm, twin, 100 watt lamp	nr	195.27			195.27
180	Recessed fitting, 1200 x 600 mm, 4 x 36 watt lamp	nr	38.29			38.29
190	Circular fitting, 2D, 16 watt lamp	nr	19.80			19.80
192	Square fitting, 2D, 28 watt lamp	nr	28.10			28.10
200	Anti corrosive polycarbonate batten fitting, IP65, 1200 mm, single, 36 watt lamp	nr	44.93			44.93
202	Anti corrosive polycarbonate batten fitting, IP65, 1200 mm, twin 36 watt lamp	nr	52.83			52.83
204	Anti corrosive polycarbonate batten fitting, IP65, 1500 mm, single, 58 watt lamp	nr	47.00			47.00
206	Anti corrosive polycarbonate batten fitting, IP65, 1500 mm, twin 58 watt lamp	nr	57.50			57.50
208	Anti corrosive polycarbonate batten fitting, IP65, 1800 mm, single, 70 watt lamp	nr	57.41			57.41
210	Anti corrosive polycarbonate batten fitting, IP65, 1800 mm, twin 70 watt lamp	nr	70.63			70.63

© NSR 01 Aug 2008 - 31 Jul 2009

V15

	V : ELECTRICAL SERVICES : WORKS OF ALTERATION/SMALL WORKS		Mat. £	Lab. £	Plant £	Total £
V1533	**FLUORESCENT LUMINAIRES : remove damaged luminaire, supply and fix new luminaire, lamp, diffuser, reflector or the like, connectors and final flexible connection**					
160	Bulkhead fitting, 2 x 8 watt lamp	nr	52.67	36.80		89.47
164	Bulkhead fitting, 38 watt 2D lamp	nr	35.20	36.80		72.00
168	Batten fitting, 1200 mm single 36 watt lamp	nr	45.57	42.30		87.87
169	Batten fitting, 1200 mm twin 36 watt lamp	nr	78.11	42.30		120.41
170	Batten fitting, 1500 mm single 58 watt lamp	nr	47.26	42.30		89.56
171	Batten fitting, 1500 mm twin 58 watt lamp	nr	83.03	42.30		125.33
172	Batten fitting, 1800 mm single 70 watt lamp	nr	57.72	42.30		100.02
173	Batten fitting, 1800 mm twin 70 watt lamp	nr	93.49	42.30		135.79
174	Batten fitting, 2400 mm single 100 watt lamp	nr	112.77	42.30		155.07
175	Batten fitting, 2400 mm twin 100 watt lamp	nr	195.27	42.30		237.57
180	Recessed fitting, 1200 x 600 mm, 4 x 36 watt	nr	38.29	69.80		108.09
190	Circular fitting, 2D, 16 watt lamp	nr	19.80	42.30		62.10
192	Square fitting, 2D, 28 watt lamp	nr	28.10	42.30		70.40
200	Anti corrosive polycarbonate batten fitting, IP65, 1200 mm, single, 36 watt lamp	nr	44.93	42.30		87.23
202	Anti corrosive polycarbonate batten fitting, IP65, 1200 mm, twin 36 watt lamp	nr	52.83	42.30		95.13
204	Anti corrosive polycarbonate batten fitting, IP65, 1500 mm, single, 58 watt lamp	nr	47.00	42.30		89.30
206	Anti corrosive polycarbonate batten fitting, IP65, 1500 mm, twin 58 watt lamp	nr	57.50	42.30		99.80
208	Anti corrosive polycarbonate batten fitting, IP65, 1800 mm, single, 70 watt lamp	nr	57.41	42.30		99.71
210	Anti corrosive polycarbonate batten fitting, IP65, 1800 mm, twin 70 watt lamp	nr	70.63	42.30		112.93
V1536	**FLUORESCENT LUMINAIRES : fix only fluorescent luminaire, lamp, diffuser, reflector or the like, connectors and final flexible connection**					
035	Bulkhead fitting, 2 x 8 watt lamp	nr		30.46		30.46
037	Bulkhead fitting, 2D, 38 watt lamps	nr		30.46		30.46
039	Bulkhead fitting, 16 - 38 watt 2D lamp	nr		30.46		30.46
041	Batten fitting, 1200 - 2400 mm, 36 - 125 watt lamp	nr		35.53		35.53
043	Batten fitting, 1200 - 2400 mm, 2 x 36 - 125 watt lamps	nr		35.53		35.53
061	Recessed fitting, 1200 x 600 mm, 3 x 36 watt lamp	nr		63.45		63.45
063	Recessed fitting, 1200 x 600 mm, 4 x 36 watt lamps	nr		63.45		63.45

© NSR 01 Aug 2008 - 31 Jul 2009

V16

V : ELECTRICAL SERVICES : WORKS OF ALTERATION/SMALL WORKS				Mat. £	Lab. £	Plant £	Total £
V1536							
	071	Circular fitting + 2D 16 watt lamp	nr		35.53		35.53
	073	Square fitting + 2D 28 watt lamp	nr		35.53		35.53
V1539		FLUORESCENT LUMINAIRES : supply and fix fluorescent luminaire, lamp, diffuser, reflector or the like, connectors and final flexible connection					
	035	Bulkhead fitting, 2 x 8 watt lamp	nr	52.67	30.46		83.13
	037	Bulkhead fitting, 2D, 38 watt lamps	nr	35.20	30.46		65.66
	039	Bulkhead fitting, 2D 16 watt lamp	nr	45.30	30.46		75.76
	050	Batten fitting, 1200 mm, 36 watt lamp	nr	45.57	35.53		81.10
	051	Batten fitting, 1200 mm, twin, 36 watt lamp	nr	78.11	35.53		113.64
	052	Batten fitting, 1500 mm, single, 58 watt lamp	nr	47.26	35.53		82.79
	053	Batten fitting, 1500 mm, twin, 58 watt lamp	nr	83.03	35.53		118.56
	054	Batten fitting, 1800 mm, single, 70 watt lamp	nr	57.72	35.53		93.25
	055	Batten fitting, 1800 mm, twin, watt lamp	nr	93.49	35.53		129.02
	056	Batten fitting, 2400 mm, single, 100 watt lamp	nr	112.77	35.53		148.30
	057	Batten fitting, 2400 mm, twin, 100 watt lamp	nr	195.27	35.53		230.80
	061	Recessed fitting, 1200 x 600 mm, 3 x 36 watt lamp	nr	35.58	63.45		99.03
	063	Recessed fitting, 1200 x 600 mm, 4 x 36 watt lamps	nr	38.29	63.45		101.74
	071	Circular fitting + 2D 16 watt lamp	nr	19.80	35.53		55.33
	073	Square fitting + 2D 28 watt lamp	nr	28.10	35.53		63.63
V1542		FLUORESCENT LUMINAIRES : remove damaged diffuser, supply and fix new					
	168	Batten fitting, 1200 mm, single, 36 watt lamp	nr	18.09	6.34		24.43
	169	Batten fitting, 1200 mm, twin, 36 watt lamp	nr	22.35	6.34		28.69
	170	Batten fitting, 1500 mm, single, 58 watt lamp	nr	20.09	6.34		26.43
	171	Batten fitting, 1500 mm, twin, 58 watt lamp	nr	24.51	6.34		30.85
	172	Batten fitting, 1800 mm, single, 70 watt lamp	nr	25.03	6.34		31.37
	173	Batten fitting, 1800 mm, twin, 70 watt lamp	nr	29.77	6.34		36.11
	174	Batten fitting, 2400 mm, single, 100 watt lamp	nr	32.40	6.34		38.74
	175	Batten fitting, 2400 mm, twin, 100 watt lamp	nr	41.62	6.34		47.96
V1543		FLUORESCENT LUMINAIRES : remove diffuser, clean and refix					
	168	Batten fitting, 1200 mm, single, 36 watt lamp	nr		6.77		6.77

© NSR 01 Aug 2008 - 31 Jul 2009

V17

V : ELECTRICAL SERVICES : WORKS OF ALTERATION/SMALL WORKS			Mat. £	Lab. £	Plant £	Total £
V1543						
169	Batten fitting, 1200 mm, twin, 36 watt lamp	nr		7.19		7.19
170	Batten fitting, 1500 mm, single, 58 watt lamp	nr		6.77		6.77
171	Batten fitting, 1500 mm, twin, 58 watt lamp	nr		7.19		7.19
172	Batten fitting, 1800 mm, single, 70 watt lamp	nr		7.19		7.19
173	Batten fitting, 1800 mm, twin, 70 watt lamp	nr		7.61		7.61
174	Batten fitting, 2400 mm, single, 100 watt lamp	nr		7.61		7.61
175	Batten fitting, 2400 mm, twin, 100 watt lamp	nr		8.46		8.46
V1544	**FLUORESCENT LUMINAIRES : remove diffuser, for replacement of fitting and refix**					
168	Batten fitting, 1200 mm, single, 36 watt lamp	nr		2.54		2.54
169	Batten fitting, 1200 mm, twin, 36 watt lamp	nr		2.96		2.96
170	Batten fitting, 1500 mm, single, 58 watt lamp	nr		2.54		2.54
171	Batten fitting, 1500 mm, twin, 58 watt lamp	nr		2.96		2.96
172	Batten fitting, 1800 mm, single, 70 watt lamp	nr		2.96		2.96
173	Batten fitting, 1800 mm, twin, 70 watt lamp	nr		3.38		3.38
174	Batten fitting, 2400 mm, single, 100 watt lamp	nr		3.38		3.38
175	Batten fitting, 2400 mm, twin, 100 watt lamp	nr		4.23		4.23
V1545	**DISCHARGE LUMINAIRES : remove damaged item of equipment, supply and fix new item of equipment**					
230	Lampholder, bayonet cap	nr	1.93	6.34		8.27
232	Lampholder, edison screw	nr	6.26	6.34		12.60
234	Lamp, low pressure sodium SOX, 18 - 35 watt	nr	19.95	4.23		24.18
236	Lamp, low pressure sodium SOX, 55 - 90 watt	nr	23.35	4.23		27.58
238	Lamp, low pressure sodium SOX, 135 watt	nr	36.60	4.23		40.83
239	Lamp, low pressure sodium SOXE, 26 - 36 watt	nr	16.82	4.23		21.05
241	Lamp, low pressure sodium SOXE, 66 - 91 watt	nr	20.55	4.23		24.78
243	Lamp, low pressure sodium SOXE, 131 watt	nr	22.08	4.23		26.31
246	Low pressure sodium ignitor, 50 - 70 watt	nr	46.81	10.15		56.96
248	Low pressure sodium capacitor, 18 watt	nr	8.40	10.15		18.55
250	Low pressure sodium capacitor, 35 - 135 watt	nr	8.40	10.15		18.55
254	High pressure sodium ballast, 250 watt	nr	32.75	10.15	3.85	46.75
256	High pressure mercury capacitor, 50 - 1000 watt	nr	5.10	10.15	3.85	19.10

© NSR 01 Aug 2008 - 31 Jul 2009   V18

V : ELECTRICAL SERVICES : WORKS OF ALTERATION/SMALL WORKS			Mat. £	Lab. £	Plant £	Total £
V1545						
258	High pressure mercury ballast, 50 - 1000 watt	nr	44.85	10.15	3.85	58.85
260	Lamp, high pressure sodium SONE/T, 50 - 70 watt	nr	19.85	4.23		24.08
262	Lamp, high pressure sodium SONE/T, 110 watt	nr	43.77	4.23		48.00
263	Lamp, high pressure sodium SONE/T, 150 - 250 watt	nr	35.50	4.23		39.73
265	Lamp, high pressure sodium SONE/T, 400 watt	nr	45.00	4.23		49.23
266	Lamp, high pressure sodium SONE/T, 1000 watt	nr	65.00	4.23	1.63	70.86
270	Lamp, high pressure sodium SONXLT, 50 - 70 watt	nr	97.91	4.23		102.14
272	Lamp, high pressure sodium SONXLT, 100 - 150 watt	nr	119.26	4.23		123.49
274	Lamp, high pressure sodium SONXLT, 250 watt	nr	163.05	4.23		167.28
275	Lamp, high pressure sodium SONXLT, 400 watt	nr	209.66	4.23		213.89
276	Lamp, high pressure sodium SONTD, 1000 watt	nr	473.88	4.23	1.63	479.74
350	Lamp, high pressure mercury MBF, 80 - 125 watt	nr	31.63	4.23		35.86
352	Lamp, high pressure mercury MBF, 250 watt	nr	77.65	4.23		81.88
353	Lamp, high pressure mercury MBF, 400 watt	nr	124.23	4.23		128.46
354	Lamp, high pressure mercury MBTF/R, 160 watt	nr	93.42	4.23		97.65
356	Lamp, high pressure mercury MBTF/R, 400 - 500 watt	nr	148.67	4.23		152.90
358	Lamp, high pressure mercury MBFU, 700 watt	nr	211.43	4.23		215.66
359	Lamp, high pressure mercury MBFU, 1000 watt	nr	275.64	4.23	1.63	281.50
V1548	DISCHARGE LUMINAIRES : fix only discharge luminaire, lamp, diffuser, reflector or the like, connectors and final flexible connection					
100	Low pressure sodium fitting, 18 - 135 watt lamp	nr		50.76		50.76
102	High pressure sodium fitting, 50 - 1000 watt lamp	nr		50.76		50.76
104	High pressure mercury fitting, 80 - 1000 watt lamp	nr		50.76		50.76
V1551	DISCHARGE LUMINAIRES : supply and fix discharge luminaire, lamp, diffuser, reflector or the like, connectors and final flexible connection					
280	Tungsten halogen fitting, 500 watt lamp	nr	9.55	50.76		60.31
282	Tungsten halogen fitting, 1000 watt lamp	nr	25.00	50.76		75.76
292	70 watt SON Floodlight IP65 C/W 70 watt SON lamp	nr	37.75	50.76		88.51
294	150 watt SON Floodlight IP65 C/W 150 SON watt lamp	nr	94.05	50.76		144.81
296	250 watt SON Floodlight IP65 C/W 250 watt SON lamp	nr	96.84	50.76		147.60
298	70 watt SON Wallwasher + Photocell C/W 70 watt SON lamp	nr	129.00	50.76		179.76

© NSR 01 Aug 2008 - 31 Jul 2009

V19

V : ELECTRICAL SERVICES : WORKS OF ALTERATION/SMALL WORKS				Mat. £	Lab. £	Plant £	Total £
V1551							
300	70 watt SON Floodlight IP65 C/W 70 watt SON lamp		nr	37.75	50.76		88.51
310	High pressure sodium fitting, 70 watt lamp		nr	76.78	50.76		127.54
312	High pressure sodium fitting, 150 watt lamp		nr	153.12	50.76		203.88
314	High pressure sodium fitting, 250 watt lamp		nr	178.25	50.76		229.01
316	High pressure sodium fitting, 400 watt lamp		nr	237.77	50.76		288.53
322	High pressure sodium fitting, 1000 watt lamp		nr	473.88	50.76		524.64
V1554	**DISCHARGE LUMINAIRES : remove damaged luminaire, fix only new luminaire, lamp, diffuser, reflector or the like, connectors and final flexible connection**						
300	Low pressure sodium fitting, 18 - 135 watt lamp		nr		59.22		59.22
302	High pressure sodium fitting, 50 - 1000 watt lamp		nr		59.22		59.22
304	High pressure mercury fitting, 50 - 1000 watt lamp		nr		59.22		59.22
V1557	**DISCHARGE LUMINAIRES : supply only luminaire and lamp**						
280	Tungsten halogen fitting, 500 watt lamp		nr	9.55			9.55
282	Tungsten halogen fitting, 1000 watt lamp		nr	25.00			25.00
292	70 watt SON Floodlight IP65 C/W 70 watt SON lamp		nr	37.75			37.75
294	150 watt SON Floodlight IP65 C/W 150 watt SON lamp		nr	94.05			94.05
296	250 watt SON Floodlight C/W IP65 250 watt SON lamp		nr	96.84			96.84
310	High pressure sodium fitting, 70 watt lamp		nr	76.78			76.78
312	High pressure sodium fitting, 150 watt lamp		nr	153.12			153.12
314	High pressure sodium fitting, 250 watt lamp		nr	178.25			178.25
316	High pressure sodium fitting, 400 watt lamp		nr	237.77			237.77
322	High pressure sodium fitting, 1000 watt lamp		nr	473.88			473.88
V1560	**DISCHARGE LUMINAIRES : remove damaged luminaire, supply and fix new luminaire, lamp, diffuser, reflector or the like, connectors and final flexible connection**						
280	Tungsten halogen fitting, 500 watt lamp		nr	9.55	25.38		34.93
282	Tungsten halogen fitting, 1000 watt lamp		nr	25.00	25.38		50.38
292	70 watt SON Floodlight C/W 70 watt SON lamp		nr	37.75	59.22		96.97
294	150 watt SON Floodlight C/W 150 watt SON lamp		nr	94.05	59.22		153.27
296	250 watt SON Floodlight C/W 250 watt SON lamp		nr	96.84	59.22		156.06
298	70 watt SON Wallwasher + photocell C/W 70 watt SON lamp		nr	129.00	59.22		188.22
310	High pressure sodium fitting, 70 watt lamp		nr	76.78	59.22		136.00

© NSR 01 Aug 2008 - 31 Jul 2009   V20

V : ELECTRICAL SERVICES : WORKS OF ALTERATION/SMALL WORKS			Mat. £	Lab. £	Plant £	Total £
V1560						
312	High pressure sodium fitting, 150 watt lamp	nr	153.12	59.22		212.34
314	High pressure sodium fitting, 250 watt lamp	nr	178.25	59.22		237.47
316	High pressure sodium fitting, 400 watt lamp	nr	237.77	59.22		296.99
322	High pressure sodium fitting, 1000 watt lamp	nr	473.88	59.22		533.10
V1563	EMERGENCY LUMINAIRES : remove damaged luminaire, fix only new luminaire, lamp, diffuser, reflector or the like, connectors and final flexible connection					
375	Self-contained fitting, 200 - 300 mm, 8 - 12 watt lamp	nr		36.80		36.80
377	Self-contained fitting, 200 - 300 mm, 2 x 8 - 12 watt lamps	nr		36.80		36.80
386	Self-contained bulkhead fitting, 16 - 38 watt 2D lamp	nr		36.80		36.80
420	Self contained fitting, 13 watt lamp	nr		36.80		36.80
V1566	EMERGENCY LUMINAIRES : supply only luminaire and lamp					
375	Self-contained fitting, 200 - 300 mm, 8 - 12 watt lamp	nr	103.73			103.73
377	Self-contained fitting, 200 - 300 mm, 2 x 8 - 18 watt lamps	nr	73.68			73.68
386	Self-contained bulkhead fitting, 16 - 38 watt 2D lamp	nr	142.90			142.90
420	Self contained fitting, 2 x 10 watt lamp	nr	249.17			249.17
V1569	EMERGENCY LUMINAIRES : remove damaged luminaire, supply and fix new luminaire, lamp, diffuser, reflector or the like, connectors and final flexible connection					
375	Self contained fitting, 200 - 300mm, 8 - 12 watt lamp	nr	103.73	36.80		140.53
377	Self contained fitting, 200 - 300mm, 2 x 8 - 12 watt lamp	nr	73.68	36.80		110.48
386	Self contained bulkhead fitting, 16 - 38 watt 2D lamp	nr	142.90	36.80		179.70
420	Self contained fitting, 2 x 10 watt lamp	nr	249.17	36.80		285.97
V1572	EMERGENCY LUMINAIRES : fix only emergency luminaire, lamp, diffuser or the like, connectors and final flexible connection					
150	Self-contained fitting, 200 - 300 mm, 8 watt lamp	nr		22.84		22.84
152	Self-contained fitting, 200 - 300 mm, 2 x 8 watt lamps	nr		22.84		22.84
160	Self-contained bulkhead fitting, 16 - 38 watt 2D lamp	nr		22.84		22.84
200	Self-contained fitting, 13 watt lamp	nr		22.84		22.84
V1575	EMERGENCY LUMINAIRES : supply and fix emergency luminaire, lamp, diffuser or the like, connectors and final flexible connection					
150	Self-contained fitting, 200 - 300 mm, 8 watt lamp (Zenith)	nr	103.73	22.84		126.57

© NSR 01 Aug 2008 - 31 Jul 2009   V21

	V : ELECTRICAL SERVICES : WORKS OF ALTERATION/SMALL WORKS		Mat. £	Lab. £	Plant £	Total £
V1575						
151	Self-contained fitting, 200 - 300 mm, 8 watt lamp (Classic 8)	nr	20.83	22.84		43.67
152	Self-contained fitting, 200 - 300 mm, 2 x 18 watt lamps	nr	73.68	22.84		96.52
160	Self-contained bulkhead fitting, 16 - 38 watt 2D lamp	nr	142.90	22.84		165.74
200	Self-contained fitting, 2 x 10 watt	nr	249.17	22.84		272.01
V1578	EMERGENCY LUMINAIRES : supply and fix emergency pack to light fitting					
370	2 cell universal spare battery	nr	24.30	6.34		30.64
372	3 cell universal spare battery	nr	49.95	8.46		58.41
374	4 cell universal spare battery	nr	66.07	10.57		76.64
376	6 cell universal spare battery	nr	60.75	12.69		73.44
400	Non-maintained conversion kit, 3 hour duration, 8 watt fluorescent lamp	nr	20.83	25.38		46.21
	V17 : LIGHTING INSTALLATION : ANCILLARIES AND EQUIPMENT					
V1700	LIGHTING SWITCHES : remove damaged ancillary item, supply and fix new ancillary item					
100	5/6 Amp, one gang, moulded plastic plate switch	nr	2.23	6.34		8.57
102	5/6 Amp, one gang, two way, moulded plastic plate switch	nr	2.65	11.42		14.07
104	5/6 Amp, one gang, intermediate, moulded plastic plate switch	nr	8.09	11.42		19.51
106	5/6 Amp, two gang, two way, moulded plastic plate switch	nr	4.76	11.42		16.18
108	5/6 Amp, three gang, two way, moulded plastic plate	nr	6.98	14.38		21.36
110	5/6 Amp, four gang, two way, moulded plastic plate switch	nr	9.32	15.23		24.55
112	5/6 Amp, six gang, two way, plate switch	nr	19.94	16.92		36.86
117	20 Amp, one way, gridswitch system moulded plastic switch	nr	2.93	5.92		8.85
119	20 Amp, two way, gridswitch system moulded plastic switch	nr	3.81	6.34		10.15
121	20 Amp, intermediate, gridswitch system moulded plastic switch	nr	7.12	6.34		13.46
123	20 Amp, one way DP, gridswitch system moulded plastic switch	nr	4.29	6.34		10.63
125	20 Amp, two way and off, gridswitch system moulded plastic switch	nr	4.85	6.34		11.19
127	20 Amp, two way, gridswitch system moulded plastic secret key switch	nr	6.68	6.34		13.02
129	20 Amp, one way DP, gridswitch system moulded plastic secret key switch	nr	6.73	6.34		13.07
131	20 Amp, two way, gridswitch system moulded plastic secret key switch	nr	6.68	6.34		13.02

© NSR 01 Aug 2008 - 31 Jul 2009   V22

	V : ELECTRICAL SERVICES : WORKS OF ALTERATION/SMALL WORKS		Mat. £	Lab. £	Plant £	Total £
V1700						
135	6 Amp, two way, gridswitch system moulded plastic push to make or break switch	nr	4.92	6.34		11.26
137	5 Amp, gridswitch system moulded plastic push to make or break switch, marked with bell symbol	nr	5.08	6.34		11.42
140	20 Amp, one way DP, gridswitch system moulded plastic switch	nr	7.04	2.11		9.15
143	Gridswitch system moulded plastic blank insert	nr	1.52	2.11		3.63
146	13 Amp gridswitch system moulded plastic fuse unit	nr	8.27	2.11		10.38
V1710	**LIGHTING SWITCHES : supply and fix plate switch, white moulded plastic plate, surface or flush mounting backbox**					
101	5/6 Amp, one gang, one way	nr	2.23	6.34		8.57
103	5/6 Amp, one gang, two way	nr	2.65	7.19		9.84
105	5/6 Amp, one gang, intermediate	nr	8.09	8.88		16.97
107	5/6 Amp, two gang, two way	nr	4.76	7.61		12.37
109	5/6 Amp, three gang, two way	nr	6.98	9.73		16.71
111	5/6 Amp, four gang, two way	nr	9.32	10.15		19.47
113	5/6 Amp, six gang, two way	nr	19.94	15.23		35.17
V1720	**LIGHTING SWITCHES : supply and fix gridswitch, white moulded plastic plate, surface or flush mounting backbox, flexible PVC insulated earth continuity conductor between box and yoke**					
120	20 Amp, one gang, one way	nr	2.93	10.57		13.50
122	20 Amp, two gang, one way	nr	3.81	14.38		18.19
124	20 Amp, three gang, one way	nr	7.61	19.88		27.49
126	20 Amp, four gang, one way	nr	7.61	23.69		31.30
128	20 Amp, six gang, one way	nr	13.95	36.38		50.33
130	Extra for two way switch in place of one way switch	nr	3.96	4.23		8.19
132	Extra for intermediate switch in place of one way switch	nr	7.34	4.23		11.57
134	Extra for two way and off switch in place of one way switch	nr	5.35	4.23		9.58
136	Extra for one way secret key switch in place of one way switch	nr	7.17	5.08		12.25
138	Extra for two way secret key switch in place of one way switch	nr	6.29	4.23		10.52
140	Extra for DP secret key switch in place of one way switch	nr	7.17	5.08		12.25
142	5 Amp, two way push to make or break switch	nr	4.92	5.08		10.00
144	10 Amp, two way push to make or break switch, marked with bell symbol	nr	5.08	5.08		10.16

© NSR 01 Aug 2008 - 31 Jul 2009   V23

V : ELECTRICAL SERVICES : WORKS OF ALTERATION/SMALL WORKS			Mat. £	Lab. £	Plant £	Total £
V1730	**LIGHTING SWITCHES : supply and fix gridswitch, aluminium or enamelled finished metalclad plate, surface mounting, pressed steel backbox, flexible PVC insulated earth continuity conductor between box and yoke**					
120	20 Amp, one gang, one way	nr	7.08	10.15		17.23
122	20 Amp, two gang, one way	nr	14.16	11.42		25.58
124	20 Amp, three gang, one way	nr	21.24	12.69		33.93
126	20 Amp, four gang, one way	nr	28.32	13.96		42.28
128	20 Amp, six gang, one way	nr	42.48	15.23		57.71
130	20 Amp, one gang, two way	nr	7.89	10.57		18.46
132	20 Amp, intermediate switch	nr	12.96	10.57		23.53
136	20 Amp, DP, one way secret key	nr	6.80	10.57		17.37
138	20 Amp, DP, two way secret key	nr	7.26	10.57		17.83
V1740	**LIGHTING SWITCHES : supply and fix ceiling switch, pull cord, moulded plastic, pattress**					
220	5/6 Amp, one way	nr	4.79	9.31		14.10
222	5/6 Amp, two way	nr	5.73	9.31		15.04
224	5/6 Amp, retractive	nr	10.86	9.31		20.17
226	16 Amp, two way	nr	9.60	9.31		18.91
228	16 Amp, DP	nr	11.53	9.31		20.84
230	Plug-in ceiling roses, moulded plastic	nr	9.55	12.69		22.24
232	Three pin plug and socket, 2 amp, 240 volt	nr	10.54	8.04		18.58
V1750	**LIGHTING SWITCHES : supply and fix dimmer switch, white moulded plastic, surface or flush mounting back box**					
240	400 Watt, one gang, one way, push	nr	14.20	7.19		21.39
242	250 Watt, two gang, two way, push	nr	22.83	8.04		30.87
V1760	**LIGHTING CONTACTORS : remove damaged contactor, supply and fix new contactor, including connections**					
720	3 Pole, 24/240 volt, 9 amp	nr	16.20	25.38		41.58
722	3 Pole, 24/240 volt, 17 amp	nr	20.50	25.38		45.88
724	3 Pole, 24/240 volt, 20 amp	nr	31.78	25.38		57.16
726	3 Pole, 24/240 volt, 32 amp	nr	41.07	25.38		66.45
728	3 Pole, 24/240 volt, 37 amp	nr	52.95	25.38		78.33

© NSR 01 Aug 2008 - 31 Jul 2009   V24

	V : ELECTRICAL SERVICES : WORKS OF ALTERATION/SMALL WORKS		Mat. £	Lab. £	Plant £	Total £
V1770	LIGHTING CONTACTORS : supply and fitting in enclosure, connections					
721	2 contacts, 24/240 volt, 25 amp	nr	36.12	19.04		55.16
723	3 contacts, 24/240 volt, 25 amp	nr	30.02	19.04		49.06
725	4 contacts, 24/240 volt, 25 amp	nr	32.21	19.04		51.25
727	3 contacts, 24/240 volt, 63 amp	nr	69.57	19.04		88.61
729	4 contacts, 24/240 volt, 63 amp	nr	89.90	19.04		108.94
V1780	LIGHTING CONTROLS : ancillary equipment - remove damaged equipment, supply and fix new					
900	Photocell	nr	113.60	12.69		126.29
	V25 : POWER INSTALLATION					
V2500	DP SWITCHES : remove damaged switch, supply and fix new switch					
020	DP, 20 Amp, 240 volt, moulded plastic switch, flex outlet, neon indicator	nr	15.90	24.53		40.43
022	DP, 32 Amp, 240 volt, moulded plastic switch, neon indicator	nr	17.27	22.00		39.27
024	DP, 45 - 50 Amp, 240 volt, moulded plastic switch, neon indicator	nr	16.28	25.80		42.08
026	DP, 45 Amp, 240 volt, moulded plastic cooker control unit, integral 13 amp switched socket outlet, neon indicator	nr	35.02	29.61		64.63
028	DP, 20 Amp, 240 volt, aluminium or enamelled finished metalclad switch, flex outlet, neon indicator	nr	18.54	19.04		37.58
V2505	DP SWITCHES : supply and fit switch, moulded plastic plate, surface or flush mounting, backbox, flexible PVC insulated earth continuity conductor between box and plate					
040	20 Amp, DP, 240 volt, neon indicator, cord outlet	nr	15.90	15.23		31.13
042	20 Amp, DP, 240 volt, neon indicator, cord outlet, engraved WATER HEATER	nr	11.99	15.23		27.22
044	30/32 Amp, DP, 240 volt, neon indicator	nr	17.27	14.80		32.07
046	45 Amp, DP, 240 volt, neon indicator	nr	16.28	15.23		31.51
048	50 Amp, DP, 240 volt, neon indicator	nr	16.76	15.23		31.99
V2510	COOKER CONTROL UNIT : supply and fit switch, cooker control unit, moulded plastic plate, surface or flush mounting, backbox, flexible PVC insulated earth continuity conductor between box and plate					
052	45 Amp, DP, integral 13 amp switched socket outlet, neon indicator	nr	35.02	15.23		50.25

© NSR 01 Aug 2008 - 31 Jul 2009　　　　　　　　　　　　　　　　　　　V25

	V : ELECTRICAL SERVICES : WORKS OF ALTERATION/SMALL WORKS		Mat. £	Lab. £	Plant £	Total £
V2515	COOKER CONTROL UNIT : supply and fit switche, cooker control unit, white enamelled metalclad plate, surface or flush mounting, backbox, flexible PVC insulated earth continuity conductor between box and plate					
052	45 Amp, DP, integral 13 amp switched socket outlet, neon indicator	nr	55.81	15.23		71.04
V2520	CEILING SWITCHES : supply and fit ceiling switch, pull cord, moulded plastic, surface or flush mounting backbox					
060	16 Amp DP, neon indicator	nr	12.64	11.00		23.64
062	45 Amp DP, neon indicator	nr	17.71	11.00		28.71
V2525	SOCKET OUTLETS : remove damaged socket outlet, supply and fix new socket outlet					
135	2 - 15 Amp, 240 volt, one gang, moulded plastic socket outlet	nr	12.97	8.46		21.43
136	13 Amp, 240 volt, two gang, moulded plastic socket outlet	nr	10.03	8.46		18.49
137	2 - 15 Amp, 240 volt, one gang, moulded plastic switched socket outlet	nr	3.71	8.46		12.17
139	13 Amp, 240 volt, two gang, moulded plastic switched socket outlet	nr	6.83	8.46		15.29
141	13 Amp, 240 volt, one gang, moulded plastic safety socket, RCCB protected, 10 - 30 mA tripping current	nr	79.03	8.46		87.49
143	2 - 15 Amp, 240 volt, one gang, aluminium or enamelled finished metalclad switched socket outlet	nr	17.16	8.46		25.62
145	13 Amp, 240 volt, two gang, aluminium or enamelled finished metalclad switched socket outlet	nr	30.20	8.46		38.66
147	13 Amp, 240 volt, one gang matt chrome finished metalclad safety socket, RCCB protected, 10 - 30 mA tripping current	nr	94.54	8.46		103.00
V2530	SOCKET OUTLETS : supply and fit white moulded plastic plate, surface or flush mounting, backbox, flexible PVC insulated earth continuity conductor between box and plate					
170	2 - 15 Amp, 240 volt, one gang	nr	12.97	9.73		22.70
172	2 - 15 Amp, 240 volt, one gang, switched	nr	16.56	9.73		26.29
174	13 Amp, 240 volt, one gang, switched, neon indicator	nr	10.00	9.73		19.73
175	13 Amp, 240 volt, two gang	nr	10.03	9.73		19.76
176	13 Amp, 240 volt, two gang, switched	nr	6.83	9.73		16.56
178	13 Amp, 240 volt, two gang, switched, neon indicator	nr	15.50	9.73		25.23
180	13 Amp, 240 volt, safety socket, one gang, neon indicator, RCCB protected, 10 - 30 mA tripping current	nr	69.36	9.73		79.09

© NSR 01 Aug 2008 - 31 Jul 2009

V : ELECTRICAL SERVICES : WORKS OF ALTERATION/SMALL WORKS			Mat. £	Lab. £	Plant £	Total £
V2535	SOCKET OUTLETS : supply and fit white moulded plastic plate, surface bench mounting, angled bench pressed steel backbox, flexible PVC insulated earth continuity conductor between box and plate					
182	13 Amp, one gang, switched, single sided	nr	7.50	12.69		20.19
184	13 Amp, one gang, switched, double sided	nr	15.00	12.69		27.69
186	13 Amp, two gang, switched, single sided	nr	11.96	12.69		24.65
188	13 Amp, two gang, switched, double sided	nr	23.92	12.69		36.61
V2540	SOCKET OUTLETS : supply and fit aluminium or enamelled finished metalclad plate, surface mounting, backbox, flexible PVC insulated earth continuity conductor between box and plate					
295	13 Amp, 24 volt, one gang, switched	nr	9.89	12.69		22.58
297	13 Amp, 240 volt, one gang, switched, neon indicator	nr	20.44	12.69		33.13
299	13 Amp, 240 volt, two gang, switched	nr	15.56	12.69		28.25
301	13 Amp, 240 volt, two gang, switched, neon indicator	nr	34.05	12.69		46.74
303	13 Amp, 240 volt, safety socket, one gang, neon indicator, RCCB protected, 10 - 30 mA tripping current	nr	80.69	12.69		93.38
V2545	FUSED CONNECTION UNITS : remove damaged fused connection unit, supply and fix new unit					
255	20 Amp, 240 volt, moulded plastic fused connection unit, flex outlet, neon indicator	nr	16.16	15.23		31.39
257	20 Amp, 240 volt, moulded plastic switched fused connection unit, flex outlet, neon indicator	nr	16.61	15.23		31.84
259	20 Amp, 240 volt, satin chrome finished metalclad fused connection unit, flex outlet	nr	12.56	15.23		27.79
261	20 Amp, 240 volt, satin chrome finished metalclad switched fused connection unit, flex outlet, neon indicator	nr	21.65	15.23		36.88
V2550	FUSED CONNECTION UNITS : supply and fit white moulded plastic plate, surface or flush mounting, backbox, flexible PVC insulated earth continuity conductor between box and plate					
350	13 Amp, 240 volt, fused, cord outlet	nr	7.43	12.27		19.70
352	13 Amp, 240 volt, fused, switched, neon indicator, cord outlet	nr	11.04	12.27		23.31
V2555	FUSED CONNECTION UNITS : supply and fit aluminium or enamelled finished metalclad plate, surface mounting, backbox, flexible PVC insulated earth continuity conductor between box and plate					
400	13 Amp, 240 volt, fused, cord outlet	nr	11.93	10.15		22.08
402	13 Amp, 240 volt, switched, neon indicator, cord outlet	nr	14.20	10.15		24.35

© NSR 01 Aug 2008 - 31 Jul 2009   V27

V : ELECTRICAL SERVICES : WORKS OF ALTERATION/SMALL WORKS			Mat. £	Lab. £	Plant £	Total £
V2560	FUSED CONNECTION UNITS : supply and fix fused connection unit, white moulded plastic plate, surface or flush mounting, moulded plastic backbox, flexible pvc insulated earth continuity conductor between box and plate					
200	13 Amp, 240 volt, fused, cord outlet	nr	14.99	12.27		27.26
202	13 Amp, 240 volt, fused, switched, neon indicator, cord outlet	nr	19.31	12.27		31.58
V2565	FUSED CONNECTION UNITS : supply and fix fused connection unit, aluminium or enamelled finished metalclad plate, surface mounting, pressed steel backbox, flexible pvc insulated earth continuity conductor between box and plate					
206	13 Amp, 240 volt, fused, cord outlet	nr	14.99	10.15		25.14
208	13 Amp, 240 volt, fused, switched, neon indicator, cord outlet	nr	19.31	10.15		29.46
V2570	CABLE OUTLETS : supply and fit cord outlet, moulded plastic plate, surface or flush mounting, backbox, flexible PVC insulated earth continuity conductor between box and plate					
420	20 Amp, one gang	nr	7.75	9.73		17.48
V2575	CABLE OUTLETS : supply and fit moulded plastic plate, surface or flush mounting, backbox, flexible PVC insulated earth continuity conductor between box and plate					
425	45 Amp, one gang	nr	8.22	17.34		25.56
V2580	BLANKING PLATES : supply and fit moulded plastic plate, surface or flush mounting backbox					
430	One gang	nr	3.70	2.11		5.81
432	Two gang	nr	6.88	2.11		8.99
V2585	PLUGS : supply and fit white moulded plastic, BS1363					
450	13 Amp, 3 pin, fused	nr	3.18	2.11		5.29
V2590	BACK BOXES : general items - remove damaged, supply and fix new					
700	Replace back box to single socket	nr	1.56	4.23		5.79
702	Replace back box to double socket	nr	2.96	4.23		7.19
	V30 : MECHANICAL SERVICES WIRING AND CONTROLS					
V3000	ISOLATORS : fix only isolator, designation label and phase discs					
010	SPN/DP, 10 - 32 amp	nr		12.69		12.69
012	SPN/DP, 35 - 63 amp	nr		17.77		17.77
014	SPN/DP, 80 - 100 amp	nr		20.30		20.30

© NSR 01 Aug 2008 - 31 Jul 2009   V28

V : ELECTRICAL SERVICES : WORKS OF ALTERATION/SMALL WORKS			Mat. £	Lab. £	Plant £	Total £
V3000						
016	TP/TPN, 10 - 32 amp	nr		20.30		20.30
018	TP/TPN, 35 - 63 amp	nr		25.38		25.38
V3010	**ISOLATORS : supply and fix isolator, designation label and phase discs**					
010	SPN/DP, 10 - 32 amp	nr	67.82	12.69		80.51
012	SPN/DP, 35 - 63 amp	nr	117.39	17.77		135.16
014	SPN/DP, 80 - 100 amp	nr	178.50	20.30		198.80
016	TP/TPN, 10 - 32 amp	nr	77.88	20.30		98.18
018	TP/TPN, 35 - 63 amp	nr	235.75	25.38		261.13
V3020	**MOTOR CONTROL SWITCHES : fix only motor control switch, designation label and phase discs**					
040	DP, 0.5 - 20 amp	nr		12.69		12.69
V3030	**MOTOR CONTROL SWITCHES : supply and fix motor control switch, designation label and phase discs**					
040	22 Amp, TP & N, Motor circuit switch	nr	16.31	12.69		29.00
V3040	**CONNECTIONS TO EQUIPMENT SUPPLIED AND FIXED BY OTHERS**					
055	Motor, fractional - 1 hp	nr		25.38		25.38
057	Motor, 3 - 6 hp	nr		31.72		31.72
059	Motor, 7 - 10 hp	nr		38.07		38.07
061	Motor, 11 - 15 hp	nr		43.57		43.57
063	Motorised heaters 1 - 3 kW	nr		12.69		12.69
065	Flush regulating door contact	nr		10.15		10.15
067	Flush regulating control box	nr		10.15		10.15
069	Flush regulating valve	nr		10.15		10.15
071	Boiler module	nr		25.38		25.38
073	Motorised valve	nr		10.15		10.15
075	Motorised valve, 2 way	nr		10.15		10.15
077	Motorised valve, 3 way	nr		10.15		10.15
079	Motorised valve, 4 way	nr		10.15		10.15
081	Gas control valve	nr		10.15		10.15
083	Immersion heater	nr		8.46		8.46

© NSR 01 Aug 2008 - 31 Jul 2009

V29

V : ELECTRICAL SERVICES : WORKS OF ALTERATION/SMALL WORKS			Mat. £	Lab. £	Plant £	Total £
V3040						
085	Instantaneous electric shower	nr		12.69		12.69
V3050	**FIX AND CONNECT ONLY EQUIPMENT SUPPLIED BY OTHERS**					
090	Thermostat	nr		7.61		7.61
092	Frostat	nr		7.61		7.61
094	Temperature detector	nr		7.61		7.61
096	Boiler controller	nr		16.50		16.50
V3060	**TIME SWITCHES : supply and fix wall mounted, including fixings, connections**					
600	To domestic central heating system	nr	78.29	12.69		90.98
V3062	**PROGRAMMERS AND TIMERS (SANGAMO): Supply only programmer/timer**					
050	Compact quartz time switch single channel 24hr	nr	30.57			30.57
060	Compact quartz time switch single channel 7day	nr	38.16			38.16
100	Electronic timer 3 pin 7 day	nr	171.57			171.57
110	Electronic timer 4 pin 7 day	nr	174.45			174.45
120	Q554 Form 2 Time Switch 3 pin 24hr	nr	125.38			125.38
130	Q554 Form 2 Time Switch 4 pin 24hr	nr	130.29			130.29
V3064	**PROGRAMMERS AND TIMERS (SANGAMO): Supply and fix wall mounted, including fixings, connections**					
050	Compact quartz time switch single channel 24hr	nr	30.57	24.53		55.10
060	Compact quartz time switch single channel 7day	nr	38.16	24.53		62.69
100	Electronic timer 3 pin 7 day	nr	171.57	28.34		199.91
110	Electronic timer 4 pin 7 day	nr	174.45	28.34		202.79
120	Q554 Form 2 Time Switch 3 pin 24hr	nr	125.38	28.34		153.72
130	Q554 Form 2 Time Switch 4 pin 24hr	nr	130.29	28.34		158.63
V3066	**PROGRAMMERS AND TIMERS (SANGAMO): Fix only wall mounted, including fixings, connections**					
050	Compact quartz time switch single channel 24hr	nr		24.53		24.53
060	Compact quartz time switch single channel 7day	nr		24.53		24.53
100	Electronic timer 3 pin 7 day	nr		28.34		28.34
110	Electronic timer 4 pin 7 day	nr		28.34		28.34
120	Q554 Form 2 Time Switch 3 pin 24hr	nr		28.34		28.34

© NSR 01 Aug 2008 - 31 Jul 2009

V30

V : ELECTRICAL SERVICES : WORKS OF ALTERATION/SMALL WORKS			Mat. £	Lab. £	Plant £	Total £
V3066						
130	Q554 Form 2 Time Switch 4 pin 24hr	nr		28.34		28.34
V3068	**PROGRAMMERS AND TIMERS (SANGAMO): Remove damaged equipment, Supply and fix wall mounted, including fixings, connections**					
050	Compact quartz time switch single channel 24hr	nr	30.57	28.76		59.33
060	Compact quartz time switch single channel 7day	nr	38.16	28.76		66.92
100	Electronic timer 3 pin 7 day	nr	171.57	32.57		204.14
110	Electronic timer 4 pin 7 day	nr	174.45	32.57		207.02
120	Q554 Form 2 Time Switch 3 pin 24hr	nr	125.38	32.57		157.95
130	Q554 Form 2 Time Switch 4 pin 24hr	nr	130.29	32.57		162.86
V3070	**GENERAL : fix only items**					
610	Wiring installation to domestic central heating system	nr		380.70		380.70
	V32 : MOTOR CONTROL AND CONTACTOR SYSTEMS					
V3200	**STARTERS (FREE MOUNTING) : remove damaged starter, fix only starter, designation label and phase discs**					
020	D-O-L, 0.1 - 1 kW	nr		19.04		19.04
022	D-O-L, 1.1 - 5 kW	nr		19.04		19.04
024	D-O-L, 5.1 - 10 kW	nr		28.76		28.76
028	Star-delta, 1 - 5 kW	nr		34.26		34.26
030	Star-delta, 6 - 15 kW	nr		45.68		45.68
V3204	**STARTERS (FREE MOUNTING) : fix only starter, designation label and phase discs**					
010	D-O-L, 0.1 - 1 kW	nr		12.69		12.69
012	D-O-L, 1.1 - 5 kW	nr		19.04		19.04
014	D-O-L, 5.1 - 10 kW	nr		19.04		19.04
018	Star-delta, 1 - 5 kW	nr		22.84		22.84
021	Star-delta, 6 - 15 kW	nr		22.84		22.84
V3208	**STARTERS (FREE MOUNTING) : supply only starter**					
020	D-O-L, 0.1 - 1 kW	nr	65.55			65.55
022	D-O-L, 1.1 - 5 kW	nr	67.68			67.68
024	D-O-L, 5.1 - 10 kW	nr	71.71			71.71

© NSR 01 Aug 2008 - 31 Jul 2009   V31

	V : ELECTRICAL SERVICES : WORKS OF ALTERATION/SMALL WORKS		Mat. £	Lab. £	Plant £	Total £
V3208						
028	Star-delta, 1 - 5 kW	nr	547.78			547.78
030	Star-delta, 6 - 15 kW	nr	569.52			569.52
V3212	STARTERS (FREE MOUNTING) : remove damaged starter, supply and fix starter, designation label and phase discs					
020	D-O-L, 0.1 - 1 kW,	nr	65.55	19.04		84.59
022	D-O-L, 1.1 - 5 kW,	nr	67.68	19.04		86.72
024	D-O-L, 5.1 - 10kW	nr	71.71	28.76		100.47
028	Star-delta 1 - 5 kW	nr	235.45	34.26		269.71
030	Star-delta 6 - 15 kW	nr	240.15	45.68		285.83
V3216	STARTERS (FREE MOUNTING) : supply and fix starter, designation label and phase discs					
010	D-O-L, 0.1 - 1 kW	nr	65.55	12.69		78.24
012	D-O-L, 1.1 - 5 kW	nr	67.68	19.04		86.72
014	D-O-L, 5.1 - 10 kW	nr	71.71	19.04		90.75
018	Star-delta, 1 - 5 kW	nr	235.45	22.84		258.29
021	Star-delta, 6 - 15 kW	nr	240.15	22.84		262.99
V3220	MOTOR CONTROL SWITCHES : remove damaged motor control switch, fix only new motor control switch, designation label and phase discs					
035	22 Amp, TP & N Motor Circuit switch	nr	16.31	19.04		35.35
V3224	MOTOR CONTROL SWITCHES : supply only motor control switch					
035	22 Amp TP & N Motor circuit switch	nr	16.31			16.31
V3228	MOTOR CONTROL SWITCHES : remove damaged motor control switch, supply and fix new motor control switch, designation label and phase discs					
035	22 Amp TP&N, Motor circuit switch	nr	16.31	19.04		35.35
V3232	STOP BUTTONS : remove damaged stop button, fix only new stop button, designation label and phase discs					
040	One button, mushroom head	nr		11.42		11.42
042	Two button, mushroom head, autolock	nr		13.54		13.54

© NSR 01 Aug 2008 - 31 Jul 2009   V32

V : ELECTRICAL SERVICES : WORKS OF ALTERATION/SMALL WORKS			Mat. £	Lab. £	Plant £	Total £
V3236	STOP BUTTONS : supply only stop button					
040	One button, mushroom head	nr	25.96			25.96
042	Two button, mushroom head, autolock	nr	55.60			55.60
V3240	STOP BUTTONS : remove damaged stop button, supply and fix only new stop button, designation label and phase discs					
040	One button, mushroom head	nr	25.96	11.42		37.38
042	Two button, mushroom head, autolock stop button	nr	63.00	13.54		76.54
V3244	STOP BUTTONS : supply and fix surface mounting, metalclad enclosure, flexible pvc insulated earth continuity conductor between box and plate, designation label and phase discs					
030	One button, mushroom head	nr	25.96	12.69		38.65
032	One button, mushroom head, autolock	nr	37.57	12.69		50.26
034	One button, mushroom head, push and turn to lock	nr	50.94	12.69		63.63
V3248	CONTROL PANELS : remove damaged item of equipment or ancillary, fix only new item of equipment or ancillary					
045	Lamp or neon	nr		11.42		11.42
047	Lens	nr		7.61		7.61
049	Push button	nr		11.42		11.42
051	Isolator, SP or DP, 5 - 30 amp	nr		28.76		28.76
053	Isolator, SP or DP, 30 - 60 amp	nr		28.76		28.76
055	Isolator, TPN, 20 amp	nr		28.76		28.76
057	Isolator, TPN, 32 amp	nr		34.26		34.26
059	Isolator, TPN, 100 amp	nr		38.07		38.07
061	Isolator, TPN, 125 amp	nr		38.07		38.07
064	Contactor or starter, D-O-L, 0.5 - 2 hp, 415 volt	nr		19.04		19.04
066	Contactor or starter, D-O-L, 3.4 - 7.5 hp, 415 volt	nr		28.76		28.76
068	Contactor or starter, D-O-L, 10 - 15 hp, 415 volt	nr		38.07		38.07
072	Terminals and rail, 1 - 4 way, 5 - 30 amp	nr		27.49		27.49
074	Terminals and rail, 1 - 4 way, 30 - 60 amp	nr		28.76		28.76
076	Terminals and rail, 4 - 8 way, 5 - 30 amp	nr		28.76		28.76
078	Terminals and rail, 4 - 8 way, 30 - 60 amp	nr		28.76		28.76
080	Terminals and rail, 8 - 12 way, 5 - 30 amp	nr		38.07		38.07
082	Terminals and rail, 8 - 12 way, 30 - 60 amp	nr		38.07		38.07
084	Terminals and rail, 12 - 16 way, 5 - 30 amp	nr		50.76		50.76

© NSR 01 Aug 2008 - 31 Jul 2009

V : ELECTRICAL SERVICES : WORKS OF ALTERATION/SMALL WORKS			Mat. £	Lab. £	Plant £	Total £
V3248						
086	Terminals and rail, 12 - 16 way, 30 - 60 amp	nr		50.76		50.76
088	Terminals and rail, 16 - 20 way, 5 - 30 amp	nr		50.76		50.76
090	Terminals and rail, 16 - 20 way, 30 - 60 amp	nr		50.76		50.76
092	Overload 1.0 - 1.4A	nr		12.69		12.69
094	Overload 1.3 - 5.0A	nr		12.69		12.69
096	Overload 4.4 - 8.5A	nr		12.69		12.69
098	Overload 7.5 - 19A	nr		12.69		12.69
V3252	**CONTROL PANELS : supply only control panel equipment**					
045	Lamp or neon	nr	6.88			6.88
047	Lens	nr	0.82			0.82
049	Push button	nr	36.92			36.92
051	Isolator, SP or DP, 5 - 30 amp	nr	67.82			67.82
053	Isolator, SP or DP, 30 - 60 amp	nr	83.00			83.00
055	Isolator, TPN, 20 Amp	nr	77.88			77.88
057	Isolator, TPN, 32 amp	nr	92.30			92.30
059	Isolator, TPN, 100 amp	nr	235.75			235.75
061	Isolator, TPN, 125 amp	nr	254.42			254.42
064	Contactor or starter, D-O-L, 0.5 - 2 hp, 415 volt	nr	67.68			67.68
066	Contactor or starter, D-O-L, 3.4 - 7.5 hp, 415 volt	nr	68.98			68.98
068	Contactor or starter, D-O-L, 10 - 15 hp, 415 volt	nr	74.64			74.64
072	Terminals and rail, 1 - 4 way, 5 - 30 amp	nr	7.70			7.70
074	Terminals and rail, 1 - 4 way, 30 - 60 amp	nr	6.88			6.88
076	Terminals and rail, 4 - 8 way, 5 - 30 amp	nr	13.30			13.30
078	Terminals and rail, 4 - 8 way, 30 - 60 amp	nr	11.66			11.66
080	Terminals and rail, 8 - 12 way, 5 - 30 amp	nr	18.90			18.90
082	Terminals and rail, 8 - 12 way, 30 - 60 amp	nr	16.44			16.44
084	Terminals and rail, 12 - 16 way, 5 - 30 amp	nr	26.60			26.60
086	Terminals and rail, 12 - 16 way, 30 - 60 amp	nr	23.32			23.32
088	Terminals and rail, 16 - 20 way, 5 - 30 amp	nr	32.20			32.20
090	Terminals and rail, 16 - 20 way, 30 - 60 amp	nr	28.10			28.10
092	Overload 1.0 - 1.4A	nr	23.90			23.90
094	Overload 1.3 - 5.0A	nr	24.20			24.20

© NSR 01 Aug 2008 - 31 Jul 2009   V34

V : ELECTRICAL SERVICES : WORKS OF ALTERATION/SMALL WORKS			Mat. £	Lab. £	Plant £	Total £
V3252						
096	Overload 4.4 - 8.5A	nr	27.00			27.00
098	Overload 7.5 - 19A	nr	28.60			28.60
V3256	**CONTROL PANELS : remove damaged item of equipment or ancillary, supply and fix new item of equipment or ancillary**					
045	Lamp or neon	nr	6.88	11.42		18.30
047	Lens	nr	0.82	7.61		8.43
049	Push button	nr	33.15	11.42		44.57
051	Isolator, SP or DP, 5 - 30 amp	nr	67.82	28.76		96.58
053	Isolator, SP or DP, 30 - 60 amp	nr	83.00	28.76		111.76
055	Isolator, TPN, 20 amp	nr	77.88	28.76		106.64
057	Isolator, TPN, 32 amp	nr	92.30	34.26		126.56
059	Isolator, TPN, 100 amp	nr	235.75	28.76		264.51
061	Isolator, TPN, 125 amp	nr	254.42	38.07		292.49
064	Contactor or starter, D-O-L, 0.5 - 2h.p., 415 volt	nr	67.68	19.04		86.72
066	Contactor or starter, D-O-L, 3.4 - 7.5h.p., 415 volt	nr	68.98	28.76		97.74
068	Contactor or starter, D-O-L, 10 - 15h.p., 415 volt	nr	74.64	38.07		112.71
072	Terminals and rail, 1 - 4 way, 5 - 30 amp	nr	7.70	27.49		35.19
074	Terminals and rail, 1 - 4 way, 30 - 60 amp	nr	6.88	28.76		35.64
076	Terminals and rail, 4 - 8 way, 5 - 30 amp	nr	13.30	28.76		42.06
078	Terminals and rail, 4 - 8 way, 30 - 60 amp	nr	11.66	28.76		40.42
080	Terminals and rail, 8 - 12 way, 5 - 30 amp	nr	18.90	38.07		56.97
082	Terminals and rail, 8 - 12 way, 30 - 60 amp	nr	16.44	38.07		54.51
084	Terminals and rail, 12 - 16 way, 5 - 30 amp	nr	26.60	50.76		77.36
086	Terminals and rail, 12 - 16 way, 30 - 60 amp	nr	23.32	50.76		74.08
088	Terminals and rail, 16 - 20 way, 5 - 30 amp	nr	32.20	50.76		82.96
090	Terminals and rail, 16 - 20 way, 30 - 60 amp	nr	28.10	50.76		78.86
092	Overload 1.0 - 1.4A	nr	23.90	12.69		36.59
094	Overload 1.3 - 5.0A	nr	24.20	12.69		36.89
096	Overload 4.4 - 8.5A	nr	27.00	12.69		39.69
098	Overload 7.5 - 19A	nr	28.60	12.69		41.29

© NSR 01 Aug 2008 - 31 Jul 2009   V35

V : ELECTRICAL SERVICES : WORKS OF ALTERATION/SMALL WORKS			Mat. £	Lab. £	Plant £	Total £
V3260	CONTACTORS (FREE MOUNTING) : remove damaged contactor, fix only new contactor, designation label and phase discs					
200	Three pole, 5.7 amp, 240 volt	nr		19.04		19.04
202	Three pole, 22 amp, 240 volt	nr		28.76		28.76
204	Three pole, 40 amp, 240 volt	nr		38.07		38.07
205	Three pole, 28 amp, 415 volt	nr		28.76		28.76
206	Three pole, 55 amp, 415 volt	nr		28.76		28.76
208	Three pole, 80 amp, 415 volt	nr		41.88		41.88
V3264	CONTACTORS (FREE MOUNTING) : supply only contactor					
200	Three pole, 5.7 amp, 240 volt	nr	16.20			16.20
202	Three pole, 22 amp, 240 volt	nr	29.15			29.15
204	Three pole, 40 amp, 240 amp	nr	88.20			88.20
205	Three pole, 28 amp, 415 volt	nr	43.20			43.20
206	Three pole, 55 amp, 415 volt	nr	122.12			122.12
208	Three pole, 80 amp, 415 volt	nr	187.89			187.89
V3268	CONTACTORS (FREE MOUNTING) : remove damaged contactor, supply and fix new contactor, designation label and phase discs					
200	Three pole, 5.7 amp, 240 volt	nr	16.20	19.04		35.24
202	Three pole, 22 amp, 240 volt	nr	29.15	28.76		57.91
204	Three pole, 40 amp, 240 volt	nr	88.20	38.07		126.27
205	Three pole, 28 amp, 415 volt	nr	43.20	28.76		71.96
306	Three pole, 55 amp, 415 volt	nr	122.12	28.76		150.88
308	Three pole, 80 amp, 415 volt	nr	187.89	41.88		229.77
V3272	CONTACTORS (FREE MOUNTING) : fix only contactors, designation labels and phase discs					
040	One pole, 5 - 30 amp	nr		12.69		12.69
042	Two pole, 5 - 30 amp	nr		19.04		19.04
043	Three pole, 5 - 30 amp	nr		19.04		19.04
044	Four pole, 5 - 30 amp	nr		19.04		19.04
046	Four pole, 30 - 60 amp	nr		19.04		19.04
V3276	CONTACTORS (FREE MOUNTING) : supply and fix contactor, designation label and phase discs					
040	Three pole, 5.7 amp, 240 volt	nr	16.20	12.69		28.89
042	Three pole, 22 amp, 240 volt	nr	29.15	19.04		48.19

© NSR 01 Aug 2008 - 31 Jul 2009

V : ELECTRICAL SERVICES : WORKS OF ALTERATION/SMALL WORKS			Mat. £	Lab. £	Plant £	Total £
V3276						
043	Three pole, 40 amp, 240 volt	nr	88.20	19.04		107.24
044	Three pole, 28 amp, 415 volt	nr	43.20	19.04		62.24
046	Three pole, 55 amp, 415 volt	nr	122.12	19.04		141.16
V3280	**MOTORS : remove damaged motor, fix only new motor**					
310	0.25 kWatt, three phase, 415 volt	nr		25.38		25.38
312	1.1 kWatt, three phase, 415 volt	nr		50.76		50.76
314	2.2 kWatt, Three phase, 415 volt	nr		76.14		76.14
316	3 kWatt, three phase, 415 volt	nr		25.38		25.38
318	4 kWatt, Three phase, 415 volt	nr		57.10		57.10
320	5.5 kWatt, Three phase, 415 volt	nr		63.45		63.45
322	7.5 kWatt, Three phase, 415 volt	nr		159.35		159.35
V3284	**MOTORS : supply only new motor**					
310	0.25 kWatt, three phase, 415 volt	nr	116.26			116.26
312	1.1 kWatt, three phase, 415 volt	nr	188.22			188.22
314	2.2 kWatt, Three phase, 415 volt	nr	249.41			249.41
316	3 kWatt, three phase, 415 volt	nr	290.46			290.46
318	4 kWatt, Three phase, 415 volt	nr	340.81			340.81
320	5.5 kWatt, Three phase, 415 volt	nr	466.58			466.58
322	7.5 kWatt, Three phase, 415 volt	nr	808.48			808.48
V3288	**MOTORS : remove damaged motor, supply and fix new motor**					
310	0.25 kWatt, three phase, 415 volt	nr	116.26	25.38		141.64
312	1.1 kWatt, three phase, 415 volt	nr	188.22	50.76		238.98
314	2.2 kWatt, Three phase, 415 volt	nr	249.41	76.14		325.55
316	3 kWatt, three phase, 415 volt	nr	290.46	25.38		315.84
318	4 kWatt, Three phase, 415 volt	nr	340.81	57.10		397.91
320	5.5 kWatt, Three phase, 415 volt	nr	466.58	63.45		530.03
322	7.5 kWatt, Three phase, 415 volt	nr	808.48	159.35		967.83

© NSR 01 Aug 2008 - 31 Jul 2009   V37

V : ELECTRICAL SERVICES : WORKS OF ALTERATION/SMALL WORKS			Mat. £	Lab. £	Plant £	Total £
	V37 : ELECTRIC HEATING INSTALLATION					
V3700	**TUBULAR HEATING** : supply and fix tubular heater, horticultural standard, metalclad, fixing brackets, interconnections, final flexible connection					
010	60 Watts per 300 mm length, single tier, 600 mm in length	nr	17.74	12.69		30.43
012	60 Watts per 300 mm length, single tier, 1200 mm in length	nr	23.70	27.32		51.02
014	60 Watts per 300 mm length, single tier, 1800 mm in length	nr	29.40	31.87		61.27
016	60 Watts per 300 mm length, single tier, 2400 mm in length	nr	47.40	36.42		83.82
018	60 Watts per 300 mm length, two tier, 600 mm in length	nr	35.48	25.38		60.86
020	60 Watts per 300 mm length, two tier, 1200 mm in length	nr	47.40	54.64		102.04
022	60 Watts per 300 mm length, two tier, 1800 mm in length	nr	58.80	63.74		122.54
024	60 Watts per 300 mm length, two tier, 2400 mm in length	nr	94.80	40.61		135.41
V3706	**HAND DRYERS** : fix only hand-dryer					
026	1.5 kW, 240 volt, push button start	nr		33.84		33.84
027	2 - 2.4kW, 240 volt, automatic start	nr		33.84		33.84
030	1.5 - 3 kW, volt, heavy duty type	nr		38.07		38.07
V3712	**HAND DRYERS** : supply only item					
020	1.5 kW, 240 volt, push button start. ABS cover	nr	75.95			75.95
026	2.4 kW, 240 volt, push button start	nr	193.50			193.50
027	2.0 kW, 240 volt, automatic start ABS polycarbonate cover	nr	103.50			103.50
028	2.4 kW, 240 volt, automatic start, stainless steel	nr	175.95			175.95
029	2.4 kW, 240 volt, automatic start, satin chrome	nr	336.00			336.00
030	2.4 kW, 240volt, automatic heavy duty hand + face drier	nr	193.50			193.50
032	2.4 kW, 240volt, automatic heavy duty manual	nr	414.75			414.75
034	2.4 kW, 240volt, automatic heavy duty auto	nr	347.75			347.75
040	2.4 kW, 240 volt, push button start, Heavy duty	nr	414.75			414.75
V3718	**HAND DRYERS** : supply and fix hand-dryer					
020	1.5 kW, 240 volt, push button start. ABS cover	nr	75.95	33.84		109.79
026	2.4 kW, 240 volt, push button start	nr	193.50	33.84		227.34
027	2.0 kW, 240 volt, automatic start ABS polycarbonate cover	nr	103.50	33.84		137.34
028	2.4 kW, 240 volt, automatic start, stainless steel	nr	175.95	33.84		209.79
029	2.4 kW, 240 volt, automatic start, satin chrome	nr	336.00	33.84		369.84
030	2.4 kW, 240volt, heavy duty hand + face drier	nr	193.50	38.07		231.57
© NSR 01 Aug 2008 - 31 Jul 2009						V38

V : ELECTRICAL SERVICES : WORKS OF ALTERATION/SMALL WORKS			Mat. £	Lab. £	Plant £	Total £
V3718						
032	2.4 kW, 240volt, heavy duty manual	nr	414.75	38.07		452.82
034	2.4 kW, 240volt, automatic heavy duty	nr	347.75	38.07		385.82
040	2.4 kW, 240 volt, push button start, Heavy duty	nr	414.75	38.07		452.82
V3724	**HEAT EMITTERS : remove damaged heat emitter, fix only new heat emitter, final connection**					
055	0.5 - 3 kW, Single phase, 240 volt	nr		38.07		38.07
057	0.5 - 3 kW, Three phase, 415 volt	nr		42.30		42.30
V3730	**HEAT EMITTERS : supply only heat emitter**					
058	1.7-3.4kw, single phase, 240v, storage heater	nr	221.07			221.07
059	2.0kw, single phase, 240 volt, free standing convector	nr	24.21			24.21
060	0.5 - 3 kW, Single phase, 240 volt, warm air curtain	nr	144.02			144.02
062	0.5 - 3 kW, Three phase, 415 volt, warm air curtain	nr	143.33			143.33
065	0.5 - 3 kW, Single phase, 240 volt, blower unit heater	nr	147.00			147.00
068	0.3 - 1 kW, Single phase, 240 volt, floor level convector	nr	42.00			42.00
069	1 - 2.5 kW, Single phase, 240 volt, floor level convector	nr	84.00			84.00
071	0.75 - 2.0kw panel heater	nr	81.55			81.55
V3736	**HEAT EMITTERS : remove damaged heat emitter, supply and fix new heat emitter, final connection**					
055	1.7 - 3.4kW, single phase, 240 volt, storage heater	nr	221.07	38.07		259.14
059	2.0kW, single phase, 240 volt, free standing convector	nr	24.21	38.07		62.28
060	0.5 - 3kW, single phase, 240 volt, warm air curtain	nr	144.02	38.07		182.09
062	0.5 - 3kW, three phase, 415 volt, warm air curtain	nr	143.33	42.30		185.63
065	0.5 - 3kW, single phase, 240 volt, blower unit heater	nr	147.00	38.07		185.07
068	0.3 - 1kW, single phase, 240 volt, floor level convector	nr	42.00	38.07		80.07
069	1 - 2.5kW, single phase, 240 volt, floor level convector	nr	84.00	38.07		122.07
071	0.75 - 2.0kw panel heater	nr	81.55	33.84		115.39
V3742	**HEAT EMITTERS : fix only heat emitter, final flexible connection**					
085	0.5 - 3 kW, Single phase, 240 volt	nr		25.38		25.38
087	0.5 - 3 kW, Three phase, 415 volt	nr		38.07		38.07
V3748	**HEAT EMITTERS : supply only item**					
120	1 - 3 kW, Single phase, 240 volt, electric bar fire	nr	110.63			110.63

© NSR 01 Aug 2008 - 31 Jul 2009   V39

V : ELECTRICAL SERVICES : WORKS OF ALTERATION/SMALL WORKS			Mat. £	Lab. £	Plant £	Total £
V3754	HEAT EMITTERS : supply and fix heat emitter, final flexible connection					
088	1.7 - 3.4 kw, single phase 240 volt, storage heater	nr	221.07	25.38		246.45
089	2 kw, single phase, 240 volt, free standing convector	nr	24.21	25.38		49.59
090	0.5 - 3 kw, single phase, 240 volt, warm air curtain	nr	144.02	25.38		169.40
092	0.5 - 3 kw, three phase, 415 volt, warm air curtain	nr	143.33	38.07		181.40
100	0.5 - 3 kw, single phase, 240 volt, blower unit heater	nr	147.00	25.38		172.38
110	0.3 - 1 kw, single phase, 240 volt, floor level convector	nr	42.00	25.38		67.38
111	1 - 2.5 kw, single phase, 240 volt, floor level convector	nr	84.00	25.38		109.38
120	1 - 3 kW, Single phase, 240 volt, electric bar fire	nr	110.63	25.38		136.01
125	0.75 - 2.0kw panel heater	nr	81.55	23.27		104.82
V3760	THERMOSTATS : remove damaged thermostat, fix only new thermostat					
070	Room thermostats	nr		14.80		14.80
072	Frost thermostats	nr		14.80		14.80
V3766	THERMOSTATS : supply only thermostat					
070	Room thermostats	nr	12.93			12.93
072	Frost thermostats	nr	21.00			21.00
V3772	THERMOSTATS : remove damaged thermostat, supply and fix new thermostat					
070	Room thermostat	nr	12.93	11.42		24.35
072	Frost thermostat	nr	21.00	11.42		32.42
V3778	THERMOSTATS : supply and fix air temperature detector, room thermostat, manual control					
150	10 - 40 Degrees C control range	nr	26.99	12.69		39.68
V3784	THERMOSTATS : supply and fix air temperature detector, horticultural thermostat, manual control					
155	3 - 27 Degrees C control range	nr	34.52	12.69		47.21
	V70 : EARTHING AND BONDING					
V7000	ANCILLARIES : bonding clamp					
010	Labelled SAFETY ELECTRICAL EARTH DO NOT REMOVE	nr	2.30	4.23		6.53

© NSR 01 Aug 2008 - 31 Jul 2009   V40

V : ELECTRICAL SERVICES : WORKS OF ALTERATION/SMALL WORKS			Mat. £	Lab. £	Plant £	Total £
V7025	ANCILLARIES : inspection cover					
015	Earth electrode concrete inspection cover, removable lid inscribed with designation	nr	36.82	25.38		62.20
V7050	EARTH RODS : earth rod, copper clad steel core, driving head, coupler, steel tip, cable clamp					
020	1000 mm In length	nr	7.50	6.34		13.84
022	2000 mm In length	nr	20.70	9.31		30.01
024	3000 mm In length	nr	33.90	13.11		47.01
026	4000 mm In length	nr	47.10	16.92		64.02
	V71 : LIGHTNING PROTECTION					
V7100	CONDUCTOR TAPE : fixings at 2m centres, tape clip screwed to masonry					
010	Bare copper tape, 25mm by 3mm	m	12.44	6.77		19.21
060	PVC Covered copper tape, 25mm by 3mm	m	14.67	6.77		21.44
066	Lead covered copper tape, 25mm by 3mm	m	33.63	6.77		40.40
078	Tinned copper tape, 25mm by 3mm	m	16.07	6.77		22.84
100	Flexible copper braid, 25mm by 3.5mm	m	15.77	6.77		22.54
V7125	AIR TERMINALS : fixed to masonry.					
200	Taper pointed air rod, copper, 15mm diameter, 500mm long	nr	22.74	2.11		24.85
202	Taper pointed air rod, copper, 15mm diameter, 1000mm long	nr	41.55	2.11		43.66
204	Taper pointed air rod, copper, 15mm diameter 2000mm long	nr	75.87	2.11		77.98
250	Air terminal base, copper	nr	17.90	8.46		26.36
270	Ridge saddle, copper	nr	39.42	8.46		47.88
300	Rod brackets, copper	nr	39.23	4.23		43.46
310	Rod to tape coupling, copper	nr	15.83	6.34		22.17
V7150	TEST & JUNCTION CLAMPS : fixed to masonry.					
502	Square tape clamp, conductor size 25mm by 3mm, copper	nr	7.17	2.11		9.28
	V85 : SYSTEM OF WIRING : CABLE SUPPORTS					
V8500	CONDUIT : PVC conduit, high impact, solvent welded joints, to BS4607					
010	16 mm Diameter	m	2.04	13.54		15.58
012	20 mm Diameter	m	2.25	13.54		15.79

© NSR 01 Aug 2008 - 31 Jul 2009   V41

	V : ELECTRICAL SERVICES : WORKS OF ALTERATION/SMALL WORKS		Mat. £	Lab. £	Plant £	Total £
V8500						
014	25 mm Diameter	m	3.75	15.65		19.40
V8504	CONDUIT : galvanised steel conduit, internally painted, heavy gauge, welded, screwed joints, to BS4568					
012	20 mm Diameter	m	5.37	12.69		18.06
014	25 mm Diameter	m	6.59	12.69		19.28
016	32 mm Diameter	m	10.19	14.80		24.99
V8508	CONDUIT : black enamelled steel conduit, heavy gauge, welded, screwed joints, to BS4568					
012	20 mm Diameter	m	3.98	12.69		16.67
014	25 mm Diameter	m	5.42	12.69		18.11
016	32 mm Diameter	m	8.54	14.80		23.34
V8512	CONDUIT : galvanised steel conduit, heavy gauge, welded, screwed joints, to BS4568					
012	20 mm Diameter	m	5.31	12.69		18.00
014	25 mm Diameter	m	6.59	12.69		19.28
016	32 mm Diameter	m	10.19	14.80		24.99
V8516	ANCILLARIES : conduit box, galvanised, BS4568, standard pattern					
920	Terminal box, 20 mm diameter	nr	4.89	8.46		13.35
922	Terminal box, 25 mm diameter	nr	6.78	8.88		15.66
924	Through box, 20 mm diameter	nr	5.82	8.46		14.28
926	Through box, 25 mm diameter	nr	8.09	8.88		16.97
928	Angle, tangent, 20 mm diameter	nr	12.54	10.15		22.69
930	Angle, tangent, 25 mm diameter	nr	16.92	10.15		27.07
932	Four way box, 20 mm diameter	nr	7.84	12.69		20.53
934	Four way box, 25 mm diameter	nr	11.26	12.69		23.95
936	Three way tangent, 20mm diameter	nr	12.04	11.42		23.46
938	Three way tangent, 25mm diameter	nr	18.86	11.42		30.28
940	Conduit box lids, fixing screws, BS4568 standard pattern	nr	0.56	4.23		4.79
V8520	ANCILLARIES : conduit box, black enamel, BS 4568, standard pattern					
920	Terminal box, 20 mm diameter	nr	4.56	8.46		13.02

© NSR 01 Aug 2008 - 31 Jul 2009   V42

	V : ELECTRICAL SERVICES : WORKS OF ALTERATION/SMALL WORKS		Mat. £	Lab. £	Plant £	Total £
V8520						
922	Terminal box, 25 mm diameter	nr	6.32	8.88		15.20
924	Through box, 20 mm diameter	nr	5.42	8.46		13.88
926	Through box, 25 mm diameter	nr	7.54	8.88		16.42
928	Angle, tangent, 20 mm diameter	nr	5.82	10.15		15.97
930	Angle, tangent, 25 mm diameter	nr	7.54	10.15		17.69
932	Four way box, 20 mm diameter	nr	7.57	12.69		20.26
934	Four way box, 25 mm diameter	nr	11.11	12.69		23.80
936	Three way tangent, 20mm diameter	nr	11.22	11.42		22.64
938	Three way tangent, 25mm diameter	nr	18.61	11.42		30.03
940	Conduit box lids, fixing screws, BS4568 standard pattern	nr	0.67	4.23		4.90
V8524	**ANCILLARIES : conduit box, PVC, BS 4568, standard pattern**					
920	Terminal box, 20 mm diameter	nr	3.37	8.46		11.83
922	Terminal box, 25 mm diameter	nr	5.33	8.88		14.21
924	Through box, 20 mm diameter	nr	3.76	8.46		12.22
926	Through box, 25 mm diameter	nr	5.78	8.88		14.66
928	Angle, tangent, 20 mm diameter	nr	3.76	10.15		13.91
930	Angle, tangent, 25 mm diameter	nr	5.78	10.15		15.93
932	Four way box, 20 mm diameter	nr	4.80	12.69		17.49
934	Four way box, 25 mm diameter	nr	7.22	12.69		19.91
936	Three way tangent, 20mm diameter	nr	8.79	11.42		20.21
938	Three way tangent, 25mm diameter	nr	10.97	11.42		22.39
940	Conduit box lids, fixing screws, BS4568 standard pattern	nr	0.97	4.23		5.20
V8528	**CONDUIT : galvanised conduit, heavy gauge solid drawn, screwed joints, to BS4568**					
012	20 mm Diameter	m	5.31	12.69		18.00
014	25 mm Diameter	m	6.59	12.69		19.28
016	32 mm Diameter	m	10.19	14.80		24.99
V8532	**CONDUIT : flexible metal conduit, PVC covered**					
012	20 mm Diameter	m	4.43	2.54		6.97
014	25 mm Diameter	m	6.06	3.38		9.44
016	32 mm Diameter	m	9.43	3.38		12.81

© NSR 01 Aug 2008 - 31 Jul 2009   V43

V : ELECTRICAL SERVICES : WORKS OF ALTERATION/SMALL WORKS				Mat. £	Lab. £	Plant £	Total £
V8536	CONDUIT : flexible conduit, glands, milled edge lock rings, serrated washers, drilling						
022	20 mm Diameter		m	4.71	11.42		16.13
024	25 mm Diameter		m	12.72	12.69		25.41
026	32 mm Diameter		m	21.40	15.23		36.63
V8540	TRUNKING : cable trunking, PVC, Mini-trunking, high impact, complete with snap on cover						
400	16 x 16 mm, Single compartment		m	4.25	3.81		8.06
402	25 x 16 mm, Single compartment		m	5.31	3.81		9.12
404	16 x 38 mm, Single compartment		m	6.75	4.23		10.98
406	25 x 38 mm, Single compartment		m	8.14	4.23		12.37
408	24 x 38 mm, Two compartment		m	9.56	5.08		14.64
V8544	TRUNKING : cable trunking, galvanised mild steel, minimum thickness 1.2 mm, complete with cover plate secured with captive screws						
420	50 x 50 mm, Single compartment		m	10.75	15.23		25.98
422	75 x 50 mm, Single compartment		m	14.11	16.07		30.18
424	75 x 75 mm, Single compartment		m	15.44	14.38		29.82
426	100 x 50 mm, Single compartment		m	16.59	16.92		33.51
428	100 x 100 mm, Single compartment		m	19.91	17.77		37.68
430	150 x 100 mm, Single compartment		m	27.57	20.30		47.87
432	150 x 150 mm, Single compartment		m	34.48	20.30		54.78
434	100 x 50 mm, Two compartment		m	19.42	19.46		38.88
436	100 x 100 mm, Two compartment		m	24.41	22.00		46.41
438	150 x 100 mm, Two compartment		m	66.57	22.84		89.41
440	150 x 150 mm, Two compartment		m	94.47	25.38		119.85
V8548	ADAPTABLE BOXES : adaptable box and lid, galvanised sheet steel, holes for cable and conduit entries, neoprene gasket						
945	75 x 75 x 37.5 mm		nr	8.10	5.08		13.18
949	75 x 75 x 75 mm		nr	5.86	7.61		13.47
951	100 x 100 x 37.5 mm		nr	5.36	16.92		22.28
953	100 x 100 x 50 mm		nr	5.36	16.92		22.28
955	100 x 100 x 75 mm		nr	6.97	19.04		26.01
957	100 x 100 x 100 mm		nr	12.35	20.30		32.65
961	150 x 150 x 50 mm		nr	7.64	19.04		26.68

© NSR 01 Aug 2008 - 31 Jul 2009

V : ELECTRICAL SERVICES : WORKS OF ALTERATION/SMALL WORKS			Mat. £	Lab. £	Plant £	Total £
V8548						
963	150 x 150 x 75 mm	nr	9.33	20.30		29.63
965	150 x 150 x 100 mm	nr	13.90	21.57		35.47
969	225 x 225 x 75 mm	nr	16.41	21.57		37.98
971	225 x 225 x 100 mm	nr	18.82	22.84		41.66
973	225 x 225 x 150 mm	nr	17.55	24.11		41.66
975	300 x 300 x 75 mm	nr	22.87	22.84		45.71
977	300 x 300 x 100 mm	nr	36.64	24.11		60.75
979	300 x 300 x 150 mm	nr	43.96	25.38		69.34
V8552	ADAPTABLE BOXES : adaptable box and lid, black enamel sheet steel, holes for cable and conduit entries, neoprene gasket					
947	75 x 75 x 75 mm	nr	3.60	6.34		9.94
953	100 x 100 x 50 mm	nr	4.03	16.92		20.95
955	100 x 100 x 75 mm	nr	4.03	19.04		23.07
957	100 x 100 x 100 mm	nr	7.47	20.30		27.77
961	150 x 75 x 50 mm	nr	6.01	19.04		25.05
963	150 x 100 x 75 mm	nr	4.84	20.30		25.14
965	150 x 150 x 100 mm	nr	8.36	21.57		29.93
969	225 x 225 x 75 mm	nr	9.96	21.57		31.53
971	225 x 225 x 100 mm	nr	11.32	22.84		34.16
973	300 x 150 x 175 mm	nr	10.08	24.11		34.19
977	300 x 300 x 75 mm	nr	13.74	24.11		37.85
979	300 x 300 x 100 mm	nr	26.90	25.38		52.28
V8556	CHANNEL : steel channel, galvanised, minimum thickness 1.5 mm					
550	25 mm, Single channel	m	0.33	5.08		5.41
552	38 mm, Single channel, galvanised, minimum thickness 2.5 mm	m	0.38	5.08		5.46
554	2 x 63.5 mm, Two channel	m	1.25	15.23		16.48
556	Steel channel cover strip, plastic, clip-in 41 mm wide	m	0.38	0.85		1.23
V8560	CAPPING : plastic cable capping					
600	12.5 mm Width	m	1.30	2.54		3.84
602	25 mm Width	m	1.78	2.96		4.74
604	38 mm Width	m	2.62	3.38		6.00

© NSR 01 Aug 2008 - 31 Jul 2009

V45

V : ELECTRICAL SERVICES : WORKS OF ALTERATION/SMALL WORKS			Mat. £	Lab. £	Plant £	Total £
V8564	CAPPING : steel cable capping, galvanised					
600	13 mm Width	m	0.28	5.50		5.78
602	25 mm Width	m	0.33	6.34		6.67
604	38 mm Width	m	0.38	6.77		7.15
V8568	CABLE TRAY : galvanised mild steel, minimum thickness 1.0 mm, fittings, connecting sleeves, earth straps, support and fixings					
700	50 mm Wide	m	5.33	10.15		15.48
702	75 mm Wide	m	5.75	12.27		18.02
704	100 mm Wide	m	6.56	22.84		29.40
706	150 mm Wide	m	7.56	25.38		32.94
708	225 mm Wide	m	10.77	38.07		48.84
V8572	CABLE TRAY : galvanised mild steel, minimum thickness 1.5 mm, fittings, connecting sleeves, earth straps, support and fixings					
706	150 mm Wide	m	7.71	25.38		33.09
708	225 mm Wide	m	9.23	38.07		47.30
710	300 mm Wide	m	13.33	50.76		64.09
712	450 mm Wide	m	22.62	63.45		86.07
714	600 mm Wide	m	29.35	84.60		113.95
V8576	CABLE TRAY : galvanised mild steel, minimum thickness 1.0 mm, returned flange, fittings, connecting sleeves, earth straps, supports and fixings					
704	100 mm Wide	m	21.00	22.84		43.84
706	150 mm Wide	m	23.45	25.38		48.83
708	225 mm Wide	m	29.47	38.07		67.54
710	300 mm Wide	m	46.94	50.76		97.70
V8580	CABLE TRAY : galvanised mild steel, minimum thickness 1.5 mm, returned flange, fittings, connecting sleeves, earth straps, supports and fixings					
704	100 mm Wide	m	32.39	22.84		55.23
706	150 mm Wide	m	36.00	25.38		61.38
708	225 mm Wide	m	46.74	38.07		84.81
710	300 mm Wide	m	52.96	50.76		103.72
712	450 mm Wide	m	91.61	63.45		155.06
714	600 mm Wide	m	126.36	84.60		210.96

© NSR 01 Aug 2008 - 31 Jul 2009   V46

V : ELECTRICAL SERVICES : WORKS OF ALTERATION/SMALL WORKS			Mat. £	Lab. £	Plant £	Total £
	V87 : SYSTEM OF WIRING : CABLES AND ANCILLARIES					
V8700	**ELECTRICAL CONNECTIONS** : disconnect electrical connections to item of equipment, reconnect electrical connections to item of equipment, tighten all screws or re-terminate cable and re-crimp lug as required					
010	1 - 4 nr Cables	nr		14.80		14.80
012	5 - 8 nr Cables	nr		16.92		16.92
014	9 - 12 nr Cables	nr		20.30		20.30
016	13 - 16 nr Cables	nr		24.53		24.53
V8704	**PVC 450/750 VOLT** : disconnect damaged cable from item of equipment or ancillary, remove cable, rewire existing conduit or trunking circuit segment in PVC insulated 450/750 volt grade copper conductor cable, circuit segment not exceeding 5 metres in length, reconnect					
011	1.5mm2, single core	nr	6.94	25.38		32.32
013	2.5mm2, single core	nr	11.55	25.38		36.93
015	4.0mm2, single core	nr	11.95	25.38		37.33
017	6.0mm2, single core	nr	25.05	27.49		52.54
018	10 mm2, Single core	nr	32.48	29.61		62.09
020	16 mm2, Single core	nr	51.28	31.72		83.00
022	25 mm2, Single core	nr	53.10	62.39		115.49
024	35 mm2, Single core	nr	66.06	64.50		130.56
026	50 mm2, Single core	nr	96.35	64.50		160.85
V8706	**LSZH 450/750 VOLT** : disconnect damaged cable from item of equipment or ancillary, remove cable, rewire existing conduit or trunking circuit segment in LSZH insulated 450/750 volt grade copper conductor cable, circuit segment not exceeding 5 metres in length, reconnect					
011	1.5mm2, single core	nr	2.14	25.38		27.52
013	2.5mm2, single core	nr	2.92	25.38		28.30
015	4.0mm2, single core	nr	5.14	25.38		30.52
017	6.0mm2, single core	nr	7.64	27.49		35.13
018	10 mm2, Single core	nr	12.99	29.61		42.60
020	16 mm2, Single core	nr	18.37	31.72		50.09
022	25 mm2, Single core	nr	36.59	62.39		98.98
024	35 mm2, Single core	nr	47.30	64.50		111.80
026	50 mm2, Single core	nr	54.34	64.50		118.84

© NSR 01 Aug 2008 - 31 Jul 2009   V47

	V : ELECTRICAL SERVICES : WORKS OF ALTERATION/SMALL WORKS		Mat. £	Lab. £	Plant £	Total £
V8708	CABLES AND CONDUCTORS : PVC 450/750 volt, PVC insulated cable, 450/750 volt grade, colour coded, stranded copper conductors					
310	1.5 mm2, Single core	m	0.73	0.85		1.58
312	2.5 mm2, Single core	m	1.08	0.85		1.93
314	4 mm2, Single core	m	2.39	0.85		3.24
316	6 mm2, Single core	m	3.47	1.27		4.74
318	10 mm2, Single core	m	6.50	1.27		7.77
320	16 mm2, Single core	m	10.26	1.69		11.95
322	25 mm2, Single core	m	10.62	2.11		12.73
324	35 mm2, Single core	m	13.21	2.11		15.32
326	50 mm2, Single core	m	19.27	2.54		21.81
V8710	CABLES AND CONDUCTORS : 450/750 volt, 6491B LSZH insulated cable, 450/750 volt grade, stranded copper conductors					
310	1.5 mm2, Single core	m	0.43	0.85		1.28
312	2.5 mm2, Single core	m	0.58	0.85		1.43
314	4 mm2, Single core	m	1.03	0.85		1.88
316	6 mm2, Single core	m	1.53	1.27		2.80
318	10 mm2, Single core	m	2.60	1.27		3.87
320	16 mm2, Single core	m	1.53	1.69		3.22
322	25 mm2, Single core	m	7.32	2.11		9.43
324	35 mm2, Single core	m	9.46	2.11		11.57
326	50 mm2, Single core	m	10.87	2.54		13.41
V8712	CABLES AND CONDUCTORS : PVC/PVC 450/750 volt, PVC insulated and sheathed cable, 450/750 volt grade, colour coded, stranded copper conductors					
316	6 mm2, Single core	m	5.01	1.27		6.28
318	10 mm2, Single core	m	7.97	1.27		9.24
322	25 mm2, Single core	m	19.51	1.69		21.20
324	35 mm2, Single core	m	29.70	2.11		31.81
V8716	PVC/PVC 300/500 VOLT : disconnect damaged cable from item of equipment or ancillary, remove cable, rewire existing conduit, trunking or surface clipped circuit segment in PVC insulated and sheathed 300/500 volt grade copper conductor cable, circuit segment not exceeding 5 metres in length, reconnect					
011	1.5mm2, single core	nr	6.94	25.38		32.32
013	2.5mm2, single core	nr	11.55	25.38		36.93

© NSR 01 Aug 2008 - 31 Jul 2009   V48

V : ELECTRICAL SERVICES : WORKS OF ALTERATION/SMALL WORKS			Mat. £	Lab. £	Plant £	Total £
V8716						
015	4.0mm2, single core	nr	11.95	25.38		37.33
017	6.0mm2, single core	nr	25.05	27.49		52.54
018	10 mm2, Single core	nr	39.86	29.61		69.47
020	16 mm2, Single core	nr	50.91	31.72		82.63
022	25 mm2, Single core	nr	104.86	60.71		165.57
024	35 mm2, Single core	nr	148.50	64.50		213.00
048	1.5 mm2, Flat two core and earth	nr	8.77	25.38		34.15
050	2.5 mm2, Flat two core and earth	nr	12.01	25.38		37.39
052	4 mm2, Flat two core and earth	nr	40.81	25.38		66.19
054	6 mm2, Flat two core and earth	nr	48.47	27.49		75.96
056	10 mm2, Flat two core and earth	nr	79.96	29.61		109.57
058	16 mm2, Flat two core and earth	nr	127.45	31.72		159.17
060	1.5 mm2, Flat three core and earth	nr	29.17	25.38		54.55
V8718	XPLE/LSZH 300/500 VOLT : disconnect damaged cable from item of equipment or ancillary, remove cable, rewire existing conduit, trunking or surface clipped circuit segment in XPLE insulated and LSZH sheathed 300/500 volt grade copper conductor cable, circuit segment not exceeding 5 metres in length,reconnect					
048	1.5 mm2, Flat two core and earth	nr	8.60	25.38		33.98
050	2.5 mm2, Flat two core and earth	nr	11.89	25.38		37.27
052	4 mm2, Flat two core and earth	nr	35.81	25.38		61.19
054	6 mm2, Flat two core and earth	nr	51.03	27.49		78.52
056	10 mm2, Flat two core and earth	nr	91.52	29.61		121.13
058	16 mm2, Flat two core and earth	nr	133.46	31.72		165.18
060	1.0 mm2, Flat three core and earth	nr	23.96	25.38		49.34
062	1.5 mm2, Flat three core and earth	nr	30.43	25.38		55.81
V8720	CABLES AND CONDUCTORS : PVC 300 volt, PVC insulated and sheathed cable, 300 volt grade, stranded copper conductors					
340	0..5 mm2, Two core	m	1.12	1.27		2.39
V8724	CABLES AND CONDUCTORS : PVC 300/500 volt, PVC insulated and sheathed cable, 300/500 volt grade, colour coded, stranded copper conductors					
345	1.5 mm2, Two core and earth continuity conductor	m	1.75	3.81		5.56
347	1.5 mm2, Three core and earth continuity conductor	m	5.83	3.81		9.64

© NSR 01 Aug 2008 - 31 Jul 2009

V49

V : ELECTRICAL SERVICES : WORKS OF ALTERATION/SMALL WORKS				Mat. £	Lab. £	Plant £	Total £
V8724							
	349	2.5 mm2, Two core and earth continuity conductor	m	2.40	3.81		6.21
	351	4 mm2, Two core and earth continuity conductor	m	8.16	4.23		12.39
	353	6 mm2, Two core and earth continuity conductor	m	9.69	4.65		14.34
	355	10 mm2, Two core and earth continuity conductor	m	15.99	5.08		21.07
	357	16 mm2, Two core and earth continuity conductor	m	25.49	5.08		30.57
V8728		PVC/PVC FLEXIBLE CORDS 300/500 VOLT : disconnect damaged cable from item of equipment or ancillary, remove cable, rewire existing conduit, surface clipped circuit segment or flexible connection in PVC insulated and sheathed 300/500 volt grade copper conductor flexible cord, circuit segment not exceeding 2 metres in length, reconnect					
	070	1 mm2, Two core	nr	2.89	12.69		15.58
	072	1.5 mm2, Two core	nr	4.10	12.69		16.79
	074	2.5 mm2, Two core	nr	8.70	12.69		21.39
	076	1 mm2, Three core	nr	2.64	12.69		15.33
	078	1.5 mm2, Three core	nr	3.65	12.69		16.34
	080	2.5 mm2, Three core	nr	7.47	12.69		20.16
	082	1 mm2, Four core	nr	7.05	12.69		19.74
	084	1.5 mm2, Four core	nr	10.05	12.69		22.74
	086	2.5 mm2, Four core	nr	15.51	12.69		28.20
V8732		CABLES AND CONDUCTORS : PVC flexible cord 300/500 volt, PVC insulated and sheathed flexible cable, 300/500 volt grade, colour coded, stranded copper conductors					
	364	0.5 mm2, Two core	m	0.70	1.69		2.39
	366	0.75 mm2, Three core	m	1.50	1.69		3.19
	368	1.0 mm2, Three core	m	1.32	1.69		3.01
	370	1.5 mm2, Three core	m	1.83	1.69		3.52
	372	1.5 mm2, Four core	m	5.02	1.69		6.71
	374	2.5 mm2, Three core	m	3.74	1.69		5.43
	376	2.5 mm2, Four core	m	7.76	1.69		9.45

© NSR 01 Aug 2008 - 31 Jul 2009   V50

V : ELECTRICAL SERVICES : WORKS OF ALTERATION/SMALL WORKS				Mat. £	Lab. £	Plant £	Total £
V8733	LSOH FLEXIBLE CORDS 300/500 VOLT : disconnect damaged cable from item of equipment or ancillary, remove cable, rewire existing conduit, surface clipped circuit segment or flexible connection in LSOH insulated and sheathed 300/500 volt grade copper conductor flexible cord, circuit segment not exceeding 2 metres in length, reconnect						
068	0.75 mm2, Two core		nr	0.89	12.69		13.58
070	1 mm2, Two core		nr	1.05	12.69		13.74
072	1.5 mm2, Two core		nr	1.39	12.69		14.08
074	2.5 mm2, Two core		nr	1.94	12.69		14.63
075	0.75 mm2, Three core		nr	1.00	12.69		13.69
076	1 mm2, Three core		nr	1.06	12.69		13.75
078	1.5 mm2, Three core		nr	1.44	12.69		14.13
080	2.5 mm2, Three core		nr	2.24	12.69		14.93
081	0.75 mm2, Four core		nr	1.16	12.69		13.85
082	1 mm2, Four core		nr	1.32	12.69		14.01
084	1.5 mm2, Four core		nr	1.70	12.69		14.39
086	2.5 mm2, Four core		nr	2.61	12.69		15.30
V8735	CABLES AND CONDUCTORS : LSOH flexible cord 300/500 volt, LSOH insulated and sheathed flexible cable, 300/500 volt grade, colour coded, stranded copper conductors						
366	0.75 mm2, Three core		m	0.44	1.69		2.13
368	1.0 mm2, Three core		m	0.53	1.69		2.22
370	1.5 mm2, Three core		m	0.72	1.69		2.41
372	1.5 mm2, Four core		m	0.85	1.69		2.54
374	2.5 mm2, Three core		m	1.12	1.69		2.81
376	2.5 mm2, Four core		m	1.31	1.69		3.00
V8736	PVC/PVC HEAT RESISTING FLEXIBLE CORDS 300/500 VOLT : disconnect damaged cable from item of equipment or ancillary, remove cable, rewire existing conduit, surface clipped circuit segment or flexible connection in PVC insulated and sheathed heat resisting 300/500 volt grade copper conductor flexible cord, circuit segment not exceeding 2 metres in length, reconnect						
070	1 mm2, Two core		nr	6.01	12.69		18.70
072	1.5 mm2, Two core		nr	8.37	12.69		21.06
074	2.5 mm2, Two core		nr	12.54	12.69		25.23
076	1 mm2, Three core		nr	6.01	14.80		20.81
078	1.5 mm2, Three core		nr	8.37	14.80		23.17
080	2.5 mm2, Three core		nr	12.54	14.80		27.34

© NSR 01 Aug 2008 - 31 Jul 2009   V51

	V : ELECTRICAL SERVICES : WORKS OF ALTERATION/SMALL WORKS		Mat. £	Lab. £	Plant £	Total £
V8736						
082	1 mm2, Four core	nr	12.43	15.23		27.66
084	1.5 mm2, Four core	nr	15.73	15.23		30.96
086	2.5 mm2, Four core	nr	20.20	15.23		35.43
V8740	CABLES AND CONDUCTORS : PVC heat resisting flexible cord 300/500 volt, PVC insulated and sheathed heat resisting flexible cable, 300/500 volt grade, colour coded, stranded copper conductors					
370	1.5 mm2, Three core	m	4.19	1.69		5.88
372	1.5 mm2, Four core	m	7.87	1.69		9.56
374	2.5 mm2, Three core	m	6.27	1.69		7.96
376	2.5 mm2, Four core	m	10.10	1.69		11.79
V8744	CABLES AND CONDUCTORS : PVC/SWA/PVC 600/1000 volt, PVC/SWA/PVC sheathed cable, 600/1000 volt grade, colour coded, stranded copper conductors					
395	1.5 mm2, Two core	m	3.80	3.81		7.61
397	1.5 mm2, Three core	m	4.27	3.81		8.08
399	1.5 mm2, Four core	m	4.74	3.81		8.55
401	2.5 mm2, Two core	m	4.37	3.81		8.18
403	2.5 mm2, Three core	m	5.23	3.81		9.04
405	2.5 mm2, Four core	m	6.05	3.81		9.86
407	4 mm2, Two core	m	6.91	4.23		11.14
409	4 mm2, Three core	m	8.11	4.23		12.34
411	4 mm2, Four core	m	9.67	4.23		13.90
413	6 mm2, Two core	m	8.72	4.65		13.37
415	6 mm2, Three core	m	10.28	4.65		14.93
417	6 mm2, Four core	m	13.17	4.65		17.82
419	10 mm2, Two core	m	13.10	5.08		18.18
421	10 mm2, Three core	m	18.36	5.08		23.44
423	10 mm2, Four core	m	21.40	5.08		26.48
425	16 mm2, Two core	m	16.04	5.08		21.12
427	16 mm2, Three core	m	22.71	5.08		27.79
429	16 mm2, Four core	m	25.17	5.08		30.25
431	25 mm2, Two core	m	16.47	5.50		21.97
433	25 mm2, Three core	m	19.59	5.50		25.09

© NSR 01 Aug 2008 - 31 Jul 2009

V52

	V : ELECTRICAL SERVICES : WORKS OF ALTERATION/SMALL WORKS		Mat. £	Lab. £	Plant £	Total £
V8744						
435	25 mm2, Four core	m	26.76	6.34		33.10
437	35 mm2, Two core	m	22.72	5.50		28.22
439	35 mm2, Three core	m	26.52	6.34		32.86
441	35 mm2, Four core	m	35.67	6.34		42.01
V8748	MICC 500/750 VOLT : disconnect damaged cable from item of equipment or ancillary, remove cable, rewire existing circuit segment in MICC 500/750 volt grade copper conductor cable, glanding and terminating segment ends, reconnect					
110	1.5 mm2, Two core	m	9.65	50.76		60.41
112	2.5 mm2, Two core	m	11.72	50.76		62.48
114	4 mm2, Two core	m	19.83	50.76		70.59
116	6 mm2, Two core (Multi plus)	m	17.07	54.99		72.06
118	1.5 mm2, Three core	m	3.71	52.88		56.59
120	2.5 mm2, Three core	m	19.88	52.88		72.76
122	4 mm2, Three core	m	19.83	52.88		72.71
124	6 mm2, Three core (Multi plus)	m	15.74	57.10		72.84
126	1.5 mm2, Four core	m	13.56	54.99		68.55
128	2.5 mm2, Four core	m	21.24	54.99		76.23
130	4 mm2, Four core (Multi plus)	m	11.77	54.99		66.76
132	6 mm2, Four core (Multi plus)	m	16.46	57.10		73.56
V8752	CABLES AND CONDUCTORS : MICC 500/750 volt, MICC cable, 500/750 volt grade, copper conductors					
395	1.5 mm2, Two core	m	6.93	5.92		12.85
397	1.5 mm2, Three core	m	9.17	5.92		15.09
399	1.5 mm2, Four core	m	10.77	6.34		17.11
401	2.5 mm2, Two core	m	9.01	6.34		15.35
403	2.5 mm2, Three core	m	15.20	6.34		21.54
405	2.5 mm2, Four core	m	18.45	6.77		25.22
V8756	MICC/PVC 500/750 VOLT : disconnect damaged cable from item of equipment or ancillary, remove cable, rewire existing circuit segment in MICC/PVC 500/750 volt grade copper conductor cable, glanding and terminating segment ends, reconnect					
110	1.5 mm2, Two core	m	15.23	50.76		65.99
112	2.5 mm2, Two core	m	12.81	50.76		63.57

© NSR 01 Aug 2008 - 31 Jul 2009   V53

V : ELECTRICAL SERVICES : WORKS OF ALTERATION/SMALL WORKS				Mat. £	Lab. £	Plant £	Total £
V8756							
114	4 mm2, Two core		m	20.92	50.76		71.68
116	6 mm2, Two core		m	16.24	54.99		71.23
118	1.5 mm2, Three core		m	16.65	52.88		69.53
120	2.5 mm2, Three core		m	21.69	52.88		74.57
122	4 mm2, Three core		m	30.10	52.88		82.98
124	6 mm2, Three core (Multi plus)		m	17.70	57.10		74.80
126	1.5 mm2, Four core		m	7.69	54.99		62.68
128	2.5 mm2, Four core		m	25.51	54.99		80.50
130	4 mm2, Four core (Multi plus)		m	13.73	54.99		68.72
132	6 mm2, Four core (Multi plus)		m	18.42	57.10		75.52
V8760	CABLES AND CONDUCTORS : MICC/PVC 500/750 volt, MICC/PVC cable, 500/750 volt grade, colour coded oversleeving, copper conductors						
387	1.0 mm2, Two core		m	6.86	5.08		11.94
389	1.0 mm2, Three core		m	8.32	5.08		13.40
391	1.0 mm2, Four core		m	9.78	5.08		14.86
393	1.0 mm2, Seven core		m	16.64	7.61		24.25
395	1.5 mm2, Two core		m	7.62	5.92		13.54
397	1.5 mm2, Three core		m	10.16	5.92		16.08
399	1.5 mm2, Four core		m	12.03	6.34		18.37
400	1.5 mm2, Seven core		m	16.64	7.61		24.25
401	2.5 mm2, Two core		m	9.62	6.34		15.96
403	2.5 mm2, Three core		m	15.20	6.34		21.54
405	2.5 mm2, Four core		m	18.45	6.77		25.22
V8764	ANCILLARIES : stuffing glands 300/500 volt, PVC cable, 300/500 volt grade, copper conductor, locknuts, drilling						
495	20 mm Diameter		nr	0.72	0.85		1.57
497	25 mm Diameter		nr	1.08	1.27		2.35
V8768	ANCILLARIES : stuffing glands 300/500 volt, heat resisting flexible cable, 300/500 volt grade, copper conductor, locknuts, drilling						
500	20 mm Diameter		nr	2.25	0.85		3.10
502	25 mm Diameter		nr	3.03	1.27		4.30

© NSR 01 Aug 2008 - 31 Jul 2009

V54

	V : ELECTRICAL SERVICES : WORKS OF ALTERATION/SMALL WORKS		Mat. £	Lab. £	Plant £	Total £
V8772	ANCILLARIES : PVC/SWA/PVC glands 600/1000 volt, PVC/SWA/PVC sheathed cable termination glands, 600/1000 volt grade, copper conductor, colour coded oversleeving, shrouds, earth tags, drilling					
510	1.5 mm2, Two core	nr	9.97	11.42		21.39
512	1.5 mm2, Three core	nr	9.97	12.69		22.66
514	1.5 mm2, Four core	nr	9.97	13.96		23.93
516	2.5 mm2, Two core	nr	9.97	11.42		21.39
518	2.5 mm2, Three core	nr	9.97	12.69		22.66
520	2.5 mm2, Four core	nr	9.97	13.96		23.93
522	4 mm2, Two core	nr	9.97	11.42		21.39
524	4 mm2, Three core	nr	9.97	12.69		22.66
526	4 mm2, Four core	nr	9.97	13.96		23.93
528	6 mm2, Two core	nr	9.97	11.42		21.39
530	6 mm2, Three core	nr	9.97	12.69		22.66
532	6 mm2, Four core	nr	9.97	13.96		23.93
534	10 mm2, Two core	nr	9.97	11.42		21.39
536	10 mm2, Three core	nr	9.97	12.69		22.66
538	10 mm2, Four core	nr	12.50	13.96		26.46
540	16 mm2, Two core	nr	12.50	12.69		25.19
542	16 mm2, Three core	nr	12.50	13.54		26.04
544	16 mm2, Four core	nr	12.50	13.96		26.46
546	25 mm2, Two core	nr	12.50	13.96		26.46
548	25 mm2, Three core	nr	12.50	14.80		27.30
550	25 mm2, Four core	nr	14.26	15.23		29.49
552	35 mm2, Two core	nr	12.50	20.30		32.80
554	35 mm2, Three core	nr	14.26	25.38		39.64
556	35 mm2, Four core	nr	14.26	30.46		44.72
558	50 mm2, Two core	nr	12.50	32.99		45.49
560	50 mm2, Three core	nr	14.26	35.53		49.79
564	50 mm2, Four core	nr	14.26	40.61		54.87
V8776	ANCILLARIES : MICC termination glands 500/750 volt grade, cold screw pot type, copper conductor, colour coded oversleeving, earth tags, drilling					
570	1.5 mm2, Two/three/four core	nr	1.73	6.77		8.50
580	2.5 mm2, Two/three/four core	nr	4.75	6.77		11.52

© NSR 01 Aug 2008 - 31 Jul 2009   V55

V : ELECTRICAL SERVICES : WORKS OF ALTERATION/SMALL WORKS			Mat. £	Lab. £	Plant £	Total £
V8780	ANCILLARIES : MICC/PVC termination glands, 500/750 volt grade MICC/PVC sheathed cable, cold screw pot type, copper conductor, colour coded oversleeving, shrouds, earth tags, drilling					
590	1.5 - 2.5 mm2, Two, three or four core	nr	4.75	6.77		11.52
V8784	ANCILLARIES : cable lug, tinned copper, crimped or soldered to cable end					
600	1.5 mm2	nr	0.25	0.85		1.10
602	2.5 mm2	nr	0.24	0.85		1.09
604	4 mm2	nr	0.29	0.85		1.14
606	6 mm2	nr	0.29	1.69		1.98
608	10 mm2	nr	0.17	1.69		1.86
610	16 mm2	nr	0.21	1.69		1.90
612	25 mm2	nr	0.26	2.11		2.37
614	35 mm2	nr	0.78	2.11		2.89
616	50 mm2	nr	1.04	3.38		4.42
	V96 : SUNDRY ITEMS					
V9600	TESTING : testing of ancillary items including overhauling and cleaning					
100	Fuse box	nr		6.34		6.34
102	Isolator	nr		8.46		8.46
104	Fused switch and switched fuse	nr		8.46		8.46
106	Consumer unit	nr		19.04		19.04
108	Light switch	nr		6.34		6.34
110	Socket outlet and fused spur	nr		8.46		8.46
112	Immersion heater	nr		12.69		12.69
114	Cooker control unit	nr		16.92		16.92
500	Single portable appliance	nr		6.34		6.34
520	Ten portable appliances	nr		42.30		42.30

© NSR 01 Aug 2008 - 31 Jul 2009

V56

			Mat. £	Lab. £	Plant £	Total £
	W : ELECTRICAL SERVICES : WORKS OF ALTERATION/SMALL WORKS					
	W50 : FIRE ALARM SYSTEM					
W5000	**BATTERIES : remove damaged battery, fix only new battery, reconnect battery and interconnections**					
018	Lead acid cell, 2 x 12 volt	nr		25.38		25.38
020	Lead acid cell, 24 volt	nr		25.38		25.38
022	Nickel cadmium cell, 24 volt	nr		25.38		25.38
W5007	**BATTERIES : supply only new battery**					
018	Lead acid cell, 2 x 12 volt	nr	163.34			163.34
020	Lead acid cell, 24 volt	nr	62.10			62.10
022	Nickel cadmium cell, 24 volt	nr	87.75			87.75
W5014	**BATTERIES : remove damaged battery, supply and fix new battery, reconnect battery and interconnections**					
018	Lead acid cell, 2 x 12 volt	nr	163.34	25.38		188.72
020	Lead acid cell, 24 volt	nr	62.10	25.38		87.48
022	Nickel cadmium cell, 24 volt	nr	87.75	25.38		113.13
W5021	**BATTERY CHARGERS : remove damaged battery charger, fix only new battery charger, flexible connection**					
030	Battery charger	nr		50.76		50.76
W5028	**BATTERY CHARGERS : supply only new battery charger**					
030	Battery charger	nr	110.00			110.00
W5035	**BATTERY CHARGERS : remove damaged battery charger, supply and fix new battery charger, flexible connection**					
030	Battery charger	nr	110.00	50.76		160.76
W5037	**MICC 500/750 VOLT : disconnect damaged cable from item of equipment or ancillary, remove cable, rewire existing circuit segment in MICC 500/750 volt grade copper conductor cable, glanding and terminating segment ends, reconnect**					
110	1.5 mm2, Two core	m	12.36	50.76		63.12
112	2.5 mm2, Two core	m	14.44	50.76		65.20
114	4 mm2, Two core	m	25.18	50.76		75.94
116	6 mm2, Two core	m	13.50	54.99		68.49

© NSR 01 Aug 2008 - 31 Jul 2009

W1

W : ELECTRICAL SERVICES : WORKS OF ALTERATION/SMALL WORKS				Mat. £	Lab. £	Plant £	Total £
W5037							
118	1.5 mm2, Three core		m	14.75	52.88		67.63
120	2.5 mm2, Three core		m	24.56	52.88		77.44
122	4 mm2, Three core		m	25.18	52.88		78.06
124	6 mm2, Three core		m	41.57	57.10		98.67
126	1.5 mm2, Four core		m	16.35	54.99		71.34
128	2.5 mm2, Four core		m	24.03	54.99		79.02
130	4 mm2, Four core		m	41.57	54.99		96.56
132	1.5 mm2, Three core		m	41.57	57.10		98.67
W5038	CABLES AND CONDUCTORS: MICC 500/750 volt, MICC cable, 500/750 volt grade, colour coded oversleeving, copper conductors.						
395	1.5 mm2, Two core		m	6.93	5.92		12.85
397	1.5 mm2, Three core		m	9.17	5.92		15.09
399	1.5 mm2, Four core		m	10.77	6.34		17.11
401	2.5 mm2, Two core		m	9.01	6.34		15.35
403	2.5 mm2, Three core		m	15.20	6.34		21.54
405	2.5 mm2, Four core		m	18.45	6.77		25.22
W5039	MICC/PVC 500/750 VOLT : disconnect damaged cable from item of equipment or ancillary, remove cable, rewire existing circuit segment in MICC/PVC 500/750 volt grade copper conductor cable, glanding and terminating segment ends, reconnect						
110	1.5 mm2, Two core		m	8.37	50.76		59.13
112	2.5 mm2, Two core		m	8.57	50.76		59.33
114	4 mm2, Two core		m	27.36	50.76		78.12
116	6 mm2, Two core		m	14.64	54.99		69.63
118	1.5 mm2, Three core		m	13.99	52.88		66.87
120	2.5 mm2, Three core		m	14.49	52.88		67.37
122	4 mm2, Three core		m	27.44	52.88		80.32
124	6 mm2, Three core		m	61.20	52.88		114.08
126	1.5 mm2, Four core		m	25.00	54.99		79.99
128	2.5 mm2, Four core		m	32.87	54.99		87.86
130	4 mm2, Four core		m	45.50	54.99		100.49
132	6 mm2, Four core		m	45.50	57.10		102.60

© NSR 01 Aug 2008 - 31 Jul 2009   W2

	W : ELECTRICAL SERVICES : WORKS OF ALTERATION/SMALL WORKS		Mat. £	Lab. £	Plant £	Total £
W5040	CABLES AND CONDUCTORS: MICC/PVC 500/750 volt, MICC/PVC cable, 500/750 volt grade, colour coded oversleeving, copper conductors.					
387	1.0 mm2, Two core	m	6.86	5.08		11.94
389	1.0 mm2, Three core	m	8.32	5.08		13.40
391	1.0 mm2, Four core	m	9.78	5.08		14.86
393	1.0 mm2, Seven core	m	16.64	7.61		24.25
395	1.5 mm2, Two core	m	7.62	5.92		13.54
397	1.5 mm2, Three core	m	10.16	5.92		16.08
399	1.5 mm2, Four core	m	12.03	6.34		18.37
400	1.5 mm2, Seven core	m	16.64	7.61		24.25
401	2.5 mm2, Two core	m	9.62	6.34		15.96
403	2.5 mm2, Three core	m	15.20	6.34		21.54
405	2.5 mm2, Four core	m	18.45	6.77		25.22
W5042	INDICATOR PANELS : remove damaged indicator panel, fix only new indicator panel, non-addressable					
035	1 - 4 Zone	nr		152.28		152.28
037	5 - 8 Zone	nr		164.97		164.97
039	9 - 12 Zone	nr		177.66		177.66
041	13 - 16 Zone	nr		387.00		387.00
043	17 - 20 Zone	nr		409.77		409.77
045	21 - 24 Zone	nr		432.53		432.53
W5049	INDICATOR PANELS : supply only indicator panel, non-addressable					
035	1 - 4 Zone	nr	221.90			221.90
037	5 - 8 Zone	nr	431.35			431.35
039	9 - 12 Zone	nr	737.90			737.90
041	13 - 16 Zone	nr	737.90			737.90
043	17 - 20 Zone	nr	959.80			959.80
045	21 - 24 Zone	nr	1265.85			1265.85
W5056	INDICATOR PANELS : remove damaged indicator panel, supply and fix new indicator panel, non-addressable					
035	1 - 4 zone	nr	221.90	152.28		374.18
037	5 - 8 zone	nr	527.95	164.97		692.92
039	9 - 12 zone	nr	737.90	177.66		915.56

© NSR 01 Aug 2008 - 31 Jul 2009

W3

W : ELECTRICAL SERVICES : WORKS OF ALTERATION/SMALL WORKS			Mat. £	Lab. £	Plant £	Total £
W5056						
041	13 - 16 zone	nr	737.90	387.00		1124.90
043	17 - 20 zone	nr	959.80	409.77		1369.57
045	21 - 24 zone	nr	1265.85	432.53		1698.38
W5063	**ANCILLARIES : contacts, detectors and sounders, remove damaged ancillary item, supply and fix new ancillary item**					
055	Break glass contact	nr	10.70	17.34		28.04
057	Heat detector, fixed temperature/rate of rise	nr	14.25	28.76		43.01
059	Smoke detector, addressable	nr	37.42	28.76		66.18
060	Smoke detector, mains/battery, ionisation	nr	27.25	28.76		56.01
063	Smoke detector, stand alone	nr	9.75	6.34		16.09
065	Bell, 12 volt DC	nr	21.26	19.04		40.30
067	Bell, 24 volt DC	nr	21.26	19.04		40.30
069	Bell, 240 volt AC	nr	34.50	19.04		53.54
072	Siren	nr	18.85	17.34		36.19
120	Spare glass for BGC	nr	1.30	2.11		3.41
W5070	**BREAK GLASS CONTACTS : surface/ flush mounting, backbox**					
610	N-O/N-C contact	nr	8.81	12.69		21.50
W5077	**BREAK GLASS CONTACTS : weatherproof enclosure, surface mounting**					
615	N-O/N-C contact	nr	27.47	16.92		44.39
W5084	**BELLS : dome bell, pressed steel gong**					
620	102 mm Diameter	nr	19.60	12.69		32.29
622	204 mm Diameter	nr	30.00	12.69		42.69
	W60 : SECURITY					
W6000	**SECURITY: supply and fit security component**					
000	Electronic digital codelock, polished brass,	nr	219.76	27.49		247.25
002	Master Card Reader	nr	172.99	31.72		204.71
004	Slave card reader	nr	172.99	31.72		204.71
006	Electric rim strike	nr	14.95	19.04		33.99
008	Electric mortice strike	nr	17.95	27.49		45.44

© NSR 01 Aug 2008 - 31 Jul 2009   W4

W : ELECTRICAL SERVICES : WORKS OF ALTERATION/SMALL WORKS				Mat. £	Lab. £	Plant £	Total £
W6000							
	010	Carbon monoxide alarm battery powered	nr	28.05	7.19		35.24
	012	Mains heat alarm with battery back-up	nr	27.25	28.76		56.01
	014	Deaf alarm, power pack, strobe and vibrating disc	nr	130.00	46.53		176.53
	016	Ionisation smoke alarm, hush button, lithium cell	nr	26.51	18.19		44.70
	018	Optical smoke alarm, hush button, lithium cell	nr	37.42	26.65		64.07
	020	CCTV, black and white camera, 15 metre cable	nr	89.05	49.32		138.37
	022	CCTV floodlight camera, PIR operated	nr	125.60	49.32		174.92
	024	12 in monochrome monitor with metal case	nr	86.78	19.04		105.82
	026	14 in colour monitor with metal case	nr	175.05	19.04		194.09
		W65 : COMMUNICATIONS/DATA					
W6500		**DATA CABLE MANAGEMENT - supply only segregated trunking and fittings for data**					
	005	100 x 25 mm trunking	m	37.43			37.43
	010	100 x 40 mm trunking	m	16.26			16.26
	015	145 X 40 mm trunking	m	36.88			36.88
	020	100 x 50 mm trunking	m	25.98			25.98
	025	100 x 60 mm trunking	m	20.85			20.85
	030	130 x 60 mm trunking	m	24.01			24.01
	035	100 x 100 mm trunking	m	32.32			32.32
W6505		**DATA CABLE MANAGEMENT - fix only segregated trunking and fittings for data**					
	005	100 x 25 mm trunking	m		6.34		6.34
	010	100 x 40 mm trunking	m		6.34		6.34
	015	145 X 40 mm trunking	m		6.34		6.34
	020	100 x 50 mm trunking	m		7.19		7.19
	025	100 x 60 mm trunking	m		7.19		7.19
	030	130 x 60 mm trunking	m		7.61		7.61
	035	100 x 100 mm trunking	m		7.61		7.61
W6510		**DATA CABLE MANAGEMENT - supply and fix segregated trunking and fittings for data**					
	005	100 x 25 mm trunking	m	37.43	6.34		43.77
	010	100 x 40 mm trunking	m	16.26	6.34		22.60

© NSR 01 Aug 2008 - 31 Jul 2009   W5

W : ELECTRICAL SERVICES : WORKS OF ALTERATION/SMALL WORKS			Mat. £	Lab. £	Plant £	Total £
W6510						
015	145 X 40 mm trunking	m	36.88	6.34		43.22
020	100 x 50 mm trunking	m	25.98	7.19		33.17
025	100 x 60 mm trunking	m	20.85	7.19		28.04
030	130 x 60 mm trunking	m	24.01	7.61		31.62
035	100 x 100 mm trunking	m	32.32	7.61		39.93
W6515	**FLOOD WIRING: supply and fixing of flexible power and data systems**					
000	Data and power supply, 3 m flexible umbilical, for suspended floors.	nr	62.30	57.10		119.40

ELECTRICAL

BASIC PRICES: LABOUR, PLANT AND MATERIALS

BASIC PRICES : LABOUR, PLANT AND MATERIALS

LABOUR

Kango Operator	KG	Hr	£14.81
Bricklayer	BR	Hr	£17.28
Bricklayers Labourer	BL	Hr	£12.93
Scaffolder	SC	Hr	£17.28
General Labourer	GL	Hr	£12.17
Heating Engineer	HE	Hr	£20.08
Heating Engineer Apprentice	HEA	Hr	£14.65
Trained plumber (Mechanical & Electrical)	TP	Hr	£20.08
Apprentice plumber	PPA	Hr	£14.65
Thermal insulation engineer	IE	Hr	£15.85
Duct fitter	DF	Hr	£16.46
Fitters mate	FM	Hr	£15.85
Electrical (Mechanical & Electrical)	E	Hr	£25.38
Electrical Apprentice	EA	Hr	£20.15
Electrical Labourer	EM	Hr	£19.08

There has been a rationalisation of rates e.g. Heating Engineer is the same rate as Trained Plumber, Electrical (Mechanical and Electrical) is the same rate as Electrician, etc., etc.

BASIC PRICES : LABOUR, PLANT AND MATERIALS

A:ELECTRICAL SERVICES - CONTRACTORS GENERAL COST ITEMS : WORKS OF ALTERATION/SMALL WORKS/REPAIR
GENERALLY
Generally

Generally; Electric radiator	Wk	19.00
Generally; Festoon lighting	Wk	32.00
Generally; Scaffold alarm, Quad beam & passive, 50m range - Hire per week	Wk	21.00
Generally; Scaffold alarm, Quad beam & passive, 50m range - set up and 4 weeks hire	Item	234.00
Generally; Scaffold lighting - Security flood lighting - per light	Wk	3.50
Generally; Scaffold lighting - Security flood lighting - set up & 4 weeks hire	nr	50.00
Generally; Genie super lift 363kg	Day	102.00

LIGHTWEIGHT ALUMIMIUM ACCESS UNITS
Lightweight alumimium access units

Lightweight alumimium access units; Chimney scaffold to full surround of centre ridge stack	Wk	220.00
Lightweight alumimium access units; Chimney scaffold to full surround of centre ridge stack	Day	110.00
Lightweight alumimium access units; Chimney scaffold to Half of centre ridge stack	Wk	110.00
Lightweight alumimium access units; Chimney scaffold to half of centre ridge stack	Day	55.00
Lightweight alumimium access units; Compact scissor lift 7.8m	Day	216.00
Lightweight alumimium access units; Compact scissor lift 7.8m	Wk	360.00
Lightweight alumimium access units; Staircase access unit with 300 - 450mm wide platform	Wk	90.00
Lightweight alumimium access units; Staircase access unit with 300 - 450mm wide platform	Day	45.00
Lightweight alumimium access units; Staircase access unit with 600 - 675mm wide platform	Wk	90.00
Lightweight alumimium access units; Staircase access unit with 600 - 675mm wide platform	Day	45.00
Lightweight alumimium access units; Window scaffold with 450mm wide platform	Wk	90.00

© NSR 2008-2009

BASIC PRICES : LABOUR, PLANT AND MATERIALS

Lightweight alumimium access units; Window scaffold with 450mm wide platform		Day	45.00
Lightweight alumimium access units; Window scaffold with 600mm wide platform		Wk	90.00
Lightweight alumimium access units; Window scaffold with 600mm wide platform		Day	45.00

NON - MECHANICAL PLANT
Non - Mechanical Plant

Non - Mechanical Plant; Adjustable base plates per 10 No		Hr	0.13
Non - Mechanical Plant; Brick guards per 10 No		Hr	0.16
Non - Mechanical Plant; Independent tied scaffold complete (per m²)		Wk	3.58
Non - Mechanical Plant; Putlog scaffold complete (per m²)		Wk	2.71
Non - Mechanical Plant; Scaffold boards per 100 feet		Hr	0.55
Non - Mechanical Plant; Scaffold fittings per 10 No		Wk	5.22
Non - Mechanical Plant; Scaffold tubes per 100 feet		Hr	0.40
Non - Mechanical Plant; Security fencing 3.5m x 2.2m panels, concrete feet		Hr	0.18
Non - Mechanical Plant; Tarpaulin 5 x 4m £17.00/wk (rate below per m2)		Wk	0.90

SCAFFOLD TOWER ON CASTORS 2.50 X 0.85M ON PLAN
Scaffold tower on castors 2.50 x 0.85m on plan

Scaffold tower on castors 2.50 x 0.85m on plan; 3.20m total Height		Day	56.75
Scaffold tower on castors 2.50 x 0.85m on plan; 3.20m total Height		Wk	113.50
Scaffold tower on castors 2.50 x 0.85m on plan; 4.20m total Height		Day	66.50
Scaffold tower on castors 2.50 x 0.85m on plan; 4.20m total Height		Wk	133.00
Scaffold tower on castors 2.50 x 0.85m on plan; 5.20m total Height		Day	76.25
Scaffold tower on castors 2.50 x 0.85m on plan; 5.20m total Height		Wk	152.50

BASIC PRICES : LABOUR, PLANT AND MATERIALS

Scaffold tower on castors 2.50 x 0.85m on plan; 6.20m total Height	Wk	172.00
Scaffold tower on castors 2.50 x 0.85m on plan; 6.20m total height	Day	86.00

SCAFFOLD TOWER ON CASTORS 2.50 X 1.45M ON PLAN
Scaffold tower on castors 2.50 x 1.45m on plan

Scaffold tower on castors 2.50 x 1.45m on plan; 3.20m total Height	Wk	113.50
Scaffold tower on castors 2.50 x 1.45m on plan; 3.20m total Height	Day	56.75
Scaffold tower on castors 2.50 x 1.45m on plan; 4.20m total Height	Wk	133.00
Scaffold tower on castors 2.50 x 1.45m on plan; 4.20m total Height	Day	66.50
Scaffold tower on castors 2.50 x 1.45m on plan; 5.20m total Height	Day	76.25
Scaffold tower on castors 2.50 x 1.45m on plan; 5.20m total Height	Wk	152.50
Scaffold tower on castors 2.50 x 1.45m on plan; 6.20m total Height	Wk	172.00
Scaffold tower on castors 2.50 x 1.45m on plan; 6.20m total height	Day	86.00

P:BUILDING FABRIC SUNDRIES

BRICKWORK AND BLOCKWORK
Brickwork and Blockwork

Brickwork and Blockwork; 5KVA Diesel Generator	Min	0.06
Brickwork and Blockwork; Dry Diamond Driller	Min	0.17
Brickwork and Blockwork; Kango hammer type 2500	Min	0.02
Brickwork and Blockwork; Sundry bits, cutters etc	Item	1.00

CONCRETE WORK
Concrete Work

Concrete Work; Disc cutter/angle grinder	Min	0.02
Concrete Work; Metal disc	nr	13.61

BASIC PRICES : LABOUR, PLANT AND MATERIALS

T:ELECTRICAL SERVICES : WORKS OF ALTERATION/SMALL WORKS
WATER HEATING

DOMESTIC IMMERSION HEATERS

HEATRAE SADIA GOLD DOT IMMERSION HEATER; 3KW 11 95.110.302R GOLD DOT	Item	31.57
ROD THERMOSTATS; 11" SINGLE POLE TST11	Item	7.46

INDUSTRIAL IMMERSION HEATERS

SANTON - INDUSTRIAL (INCOLOY SHEATH) 2 1/4 BSP; 6KW 16 1 OR 3 PHASE ELEC15206_MY616	Item	160.00

POINT-OF-USE WATER HEATERS

HEATRAE SADIA STREAMLINE RANGE; HEATRAE STREAMLINE 10/3 3KW 10LITRE PACK 95.010.187	Item	221.26
HEATRAE SADIA STREAMLINE RANGE; HEATRAE STREAMLINE 7/3 3KW 7LITRE PACK 95.010.183	Item	179.31

SHOWERS

HEATRAE SADIA SHOWERS; SHOWER 8.5KW 95.021.586 CAROUSEL	Item	218.28
HEATRAE SADIA SHOWERS; SHOWER 9.8KW 95.021.548 ACCOLADE 9.8KW	Item	256.11

STORAGE/CISTERN TYPE WATER HEATERS

HEATRAE SADIA RECTANGULAR FBM RANGE; 3KW 125LITRE FBM125	Item	1,214.62
HEATRAE SADIA RECTANGULAR FBM RANGE; 3KW 25LITRE FBM25	Item	720.43
HEATRAE SADIA RECTANGULAR FBM RANGE; 3KW 50LITRE FBM50	Item	892.56
HEATRAE SADIA RECTANGULAR FBM RANGE; 3KW 75LITRE FBM75	Item	1,061.29
HEATRAE SADIA RECTANGULAR FBM RANGE; 6KW 125LITRE FBM125	Item	1,290.87
ZIP-SOLECTRA STORAGE; 25 LITRE MULTIPOINT ELECTRONIC VENTED STORAGE XL25	Item	809.00
ZIP-SOLECTRA STORAGE; 50 LITRE MULTIPOINT ELECTRONIC VENTED STORAGE XL50	Item	959.00

WATER HEATER SPARES

HEATRAE SADIA SPARES; 11" THERMOSTAT 95.980.035	Item	11.75
REDRING SPARES; 3KW ELEMENT ASSEMBLY FOR WS7 WATER HEATER 94-780414	Item	40.00

BASIC PRICES : LABOUR, PLANT AND MATERIALS

U:ELECTRICAL SERVICES : WORKS OF ALTERATION/SMALL WORKS
VENTILATION

BATHROOM/TOILET EXTRACTOR FANS

NEWLEC BATHROOM / TOILET FAN; 100MM STANDARD AXIAL FAN NL880	Item	12.50

CONTROLLERS AND ACCESSORIES

VENT AXIA CONTROLLERS; ELECTRONIC CONTROLLER 10303106A	Item	110.30
XPELAIR CONTROLLERS; 21854AW GROUP CONTROLLER FOR GX/RX/WX FANS EC6H	Item	90.67

ROOF MOUNTED EXTRACTOR FANS

XPELAIR ROOF FANS; 12 ROOF FAN RX12	Item	374.69
XPELAIR ROOF FANS; 9 ROOF FAN RX9	Item	283.84

WINDOW/WALL MOUNTED EXTRACTOR FANS

XPELAIR GX RANGE WINDOW / WALL FANS; 12 AUTOMATIC SHUTTER OPERATION GX12	Item	283.84
XPELAIR GX RANGE WINDOW / WALL FANS; 9 AUTOMATIC SHUTTER OPERATION GX9	Item	181.64
XPELAIR GX RANGE WINDOW / WALL FANS; 90800AW 6 FAN. AUTOMATIC SHUTTER OPERATION GX6	Item	90.79
XPELAIR GX RANGE WINDOW / WALL FANS; WALL KIT FOR GX12 HOLE SIZE - 330MM WK12/11	Item	18.52
XPELAIR GX RANGE WINDOW / WALL FANS; WALL KIT FOR GX6 HOLE SIZE - 203MM WK6/11	Item	11.71
XPELAIR GX RANGE WINDOW / WALL FANS; WALL KIT FOR GX9 HOLE SIZE - 267MM WK9/11	Item	15.73

© NSR 2008-2009

BASIC PRICES : LABOUR, PLANT AND MATERIALS

V:ELECTRICAL SERVICES : WORKS OF ALTERATION/SMALL WORKS
CABLE & CABLE ACCESSORIES
CABLE COVERS AND ACCESSORIES

CABLE CONCRETE ACCESSORIES; EARTH ROD PIT PT205	Item	36.82

CABLE GLANDS AND GLAND ACCESSORIES

ACER CABLE GLANDS; CABLE GLAND 20MM (INDOOR/OUTDOOR TYPE) NL20CWN	Item	9.97
ACER CABLE GLANDS; CABLE GLAND 25MM (INDOOR/OUTDOOR TYPE) NL25CWN	Item	12.50
ACER CABLE GLANDS; CABLE GLAND 32MM (INDOOR/OUTDOOR TYPE) NL32CWN	Item	14.26
NYLON GLANDS AND LOCKNUTS; BLACK M20 STANDARD GLAND 403KB53	Item	0.51
NYLON GLANDS AND LOCKNUTS; BLACK M25 STANDARD GLAND 403KB55	Item	0.87
NYLON GLANDS AND LOCKNUTS; M25X1.5 LOCKNUT BLACK ML25B	Item	0.21
NYLON GLANDS AND LOCKNUTS; M25X1.5 LOCKNUT WHITE ML25W	Item	0.21
NYLON GLANDS FOR UN-ARMOURED CABLES; 20MM NYLON GLAND+NUT (INDOOR/OUTDOOR TYPE) 403AT53	Item	2.25
NYLON GLANDS FOR UN-ARMOURED CABLES; 25MM NYLON GLAND+NUT (INDOOR/OUTDOOR TYPE) 403AT55	Item	3.03

CABLE LUGS / COPPER TUBE TERMINALS

LUG TERMINALS; COPPER TUBE LUG 1.5-2.5MM^2 5MM STUD BT2C4	Item	0.25
LUG TERMINALS; COPPER TUBE LUG 1.5-2.5MM^2 6MM STUD BT2C6	Item	0.24
LUG TERMINALS; COPPER TUBE LUG 35MM^2 8MM STUD BT35C8	Item	0.78
LUG TERMINALS; COPPER TUBE LUG 4-6MM^2 6MM STUD BT6C6	Item	0.29
LUG TERMINALS; COPPER TUBE LUG 50MM^2 12MM STUD BT50C12	Item	1.04
NEWLEC COPPER TUBE TERMINALS (BULK PACKS) - NEW RANGE; 10MM COPPER TUBE LUG - 6MM STUD (BULK PACK) NL10C6X	Item	8.70
NEWLEC COPPER TUBE TERMINALS (BULK PACKS) - NEW RANGE; 16MM COPPER TUBE LUG - 6MM STUD (BULK PACK) NL16C6X	Item	10.25
NEWLEC COPPER TUBE TERMINALS (BULK PACKS) - NEW RANGE; 25MM COPPER TUBE LUG - 8MM STUD (BULK PACK) NL25C8X	Item	12.95

EARTHING / LIGHTNING PROTECTION

FURSE EARTH CLAMPS; OUTDOOR EARTH CLAMP 2X16MM TERMINALS A-D P/BRONZE NLEC16N	Item	1.55
FURSE EARTH RODS, EARTHING TAPE + ACCESSORIES; 25X3.5MM FLEXIBLE COPPER BRAID BD030	m	15.43

BASIC PRICES : LABOUR, PLANT AND MATERIALS

V:ELECTRICAL SERVICES : WORKS OF ALTERATION/SMALL WORKS
CABLE & CABLE ACCESSORIES
EARTHING / LIGHTNING PROTECTION

Description	Unit	Price
FURSE EARTH RODS, EARTHING TAPE + ACCESSORIES; 25X3MM BARE COPPER TAPE TC030	m	12.10
FURSE EARTH RODS, EARTHING TAPE + ACCESSORIES; 25X3MM LEAD COVERED COPPER TAPE TC330	m	33.30
FURSE EARTH RODS, EARTHING TAPE + ACCESSORIES; 25X3MM PVC COVERED GREEN/YELLOW COPPER TAPE TC105-FU	m	14.33
FURSE EARTH RODS, EARTHING TAPE + ACCESSORIES; 25X3MM TINNED COPPER TAPE TC230	m	15.73
FURSE EARTH RODS, EARTHING TAPE + ACCESSORIES; 3/8 EARTH ROD AND CLAMP NL780	Item	7.50
FURSE EARTH RODS, EARTHING TAPE + ACCESSORIES; 5/8 EARTH ROD 4FT NL781	Item	8.95
FURSE EARTH RODS, EARTHING TAPE + ACCESSORIES; 5/8 ROD COUPLER NL782	Item	4.25
FURSE EARTH RODS, EARTHING TAPE + ACCESSORIES; AIR TERMINAL BASE SD307	nr	17.90
FURSE EARTH RODS, EARTHING TAPE + ACCESSORIES; NON-METALLIC CLIP - BROWN CP025	Item	0.67
FURSE EARTH RODS, EARTHING TAPE + ACCESSORIES; NON-METALLIC CLIP - GREEN CP035	Item	0.67
FURSE EARTH RODS, EARTHING TAPE + ACCESSORIES; RIDGE SADDLE SD115	nr	39.42
FURSE EARTH RODS, EARTHING TAPE + ACCESSORIES; ROD BRACKETS BR105	nr	39.23
FURSE EARTH RODS, EARTHING TAPE + ACCESSORIES; ROD TO TAPE COUPLING CG600	nr	15.83
FURSE EARTH RODS, EARTHING TAPE + ACCESSORIES; SQUARE TAPE CLAMP CT105-FU	Item	7.17
FURSE EARTH RODS, EARTHING TAPE + ACCESSORIES; TAPER POINTED AIR ROD 1000MM RA225	nr	41.55
FURSE EARTH RODS, EARTHING TAPE + ACCESSORIES; TAPER POINTED AIR ROD 2000MM RA240	nr	75.87
FURSE EARTH RODS, EARTHING TAPE + ACCESSORIES; TAPER POINTED AIR ROD 500MM RA215	nr	22.74

FLEXIBLE CORDS

Description	Unit	Price
2182Y PVC INSULATED/PVC SHEATHED - TWIN CIRCULAR; 100MTR 0.5MM	1000m	704.30
2183Y PVC INSULATED/PVC SHEATHED - 3 CORE CIRCULAR; 100MTR 0.75MM	1000m	1,496.70
3093Y PVC INSULATED/PVC SHEATHED, HEAT RESISTANT 85° C - 3 CORE CIRCULAR; 100MTR 1.0MM	1000m	3,005.80
3093Y PVC INSULATED/PVC SHEATHED, HEAT RESISTANT 85° C - 3 CORE CIRCULAR; 100MTR 1.5MM	1000m	4,185.10
3093Y PVC INSULATED/PVC SHEATHED, HEAT RESISTANT 85° C - 3 CORE CIRCULAR; 100MTR 2.5MM	1000m	6,271.70

© NSR 2008-2009

BASIC PRICES : LABOUR, PLANT AND MATERIALS

V:ELECTRICAL SERVICES : WORKS OF ALTERATION/SMALL WORKS
CABLE & CABLE ACCESSORIES
FLEXIBLE CORDS

3182Y PVC INSULATED/PVC SHEATHED - 2 CORE CIRCULAR; WHITE 100MTR 1.0MM	1000m	1,442.80
3182Y PVC INSULATED/PVC SHEATHED - 2 CORE CIRCULAR; WHITE 100MTR 1.5MM	1000m	2,051.10
3182Y PVC INSULATED/PVC SHEATHED - 2 CORE CIRCULAR; WHITE 100MTR 2.5MM	1000m	4,350.80
3182Y PVC INSULATED/PVC SHEATHED - 2 CORE CIRCULAR; WHITE 100MTR 0.75MM	1000m	1,115.50
3183Y PVC INSULATED/PVC SHEATHED - 3 CORE CIRCULAR; BLACK 100MTR 1.0MM	1000m	1,317.70
3183Y PVC INSULATED/PVC SHEATHED - 3 CORE CIRCULAR; BLACK 100MTR 1.5MM	1000m	1,825.80
3183Y PVC INSULATED/PVC SHEATHED - 3 CORE CIRCULAR; BLACK 100MTR 2.5MM	1000m	3,736.70
3184TQ EPR INSULATED/HOFR SHEATHED. HEAT RESISTANT 85° C -4 CORE CIRCULAR; 100MTR 1.0MM	1000m	6,214.70
3184TQ EPR INSULATED/HOFR SHEATHED. HEAT RESISTANT 85° C -4 CORE CIRCULAR; 100MTR 2.5MM	1000m	10,099.00
3184TQ EPR INSULATED/HOFR SHEATHED. HEAT RESISTANT 85° C -4 CORE CIRCULAR; 100MTR 1.5MM	1000m	7,865.00
3184Y PVC INSULATED/PVC SHEATHED - 4 CORE CIRCULAR; BLACK 100MTR 1.0MM	1000m	3,526.30
3184Y PVC INSULATED/PVC SHEATHED - 4 CORE CIRCULAR; BLACK 100MTR 1.5MM	1000m	5,022.90
3184Y PVC INSULATED/PVC SHEATHED - 4 CORE CIRCULAR; BLACK 100MTR 2.5MM	1000m	7,755.10
LSOH CIRCULAR 2 CORE 3182B; 3182B 0.75MM 2 CORE	1000m	443.70
LSOH CIRCULAR 2 CORE 3182B; 3182B 1.0MM 2 CORE	1000m	525.00
LSOH CIRCULAR 2 CORE 3182B; 3182B 1.5MM 2 CORE	1000m	697.10
LSOH CIRCULAR 2 CORE 3182B; 3182B 2.5MM 2 CORE	1000m	969.70
LSOH CIRCULAR 3 CORE 3183B; 3183B 0.75MM 3 CORE	1000m	499.60
LSOH CIRCULAR 3 CORE 3183B; 3183B 1.0MM 3 CORE	1000m	530.30
LSOH CIRCULAR 3 CORE 3183B; 3183B 1.5MM 3 CORE	1000m	721.60
LSOH CIRCULAR 3 CORE 3183B; 3183B 2.5MM 3 CORE	1000m	1,120.90
LSOH CIRCULAR 4 CORE 3184B; 3184B 0.75MM 4 CORE	1000m	579.80
LSOH CIRCULAR 4 CORE 3184B; 3184B 1.0MM 4 CORE	1000m	660.30
LSOH CIRCULAR 4 CORE 3184B; 3184B 1.5MM 4 CORE	1000m	852.30
LSOH CIRCULAR 4 CORE 3184B; 3184B 2.5MM 4 CORE	1000m	1,306.80

GENERAL WIRING CABLES

6181YH PVC INSULATED/PVC SHEATHED. SINGLE CORE; GREY/BLUE 100MTR 1.5MM	1000m	1,387.00

BASIC PRICES : LABOUR, PLANT AND MATERIALS

V:ELECTRICAL SERVICES : WORKS OF ALTERATION/SMALL WORKS
CABLE & CABLE ACCESSORIES

GENERAL WIRING CABLES

Description	Unit	Price
6181YH PVC INSULATED/PVC SHEATHED. SINGLE CORE; GREY/BLUE 100MTR 10MM	1000m	7,971.00
6181YH PVC INSULATED/PVC SHEATHED. SINGLE CORE; GREY/BLUE 100MTR 16MM	1000m	10,182.30
6181YH PVC INSULATED/PVC SHEATHED. SINGLE CORE; GREY/BLUE 100MTR 2.5MM	1000m	2,310.80
6181YH PVC INSULATED/PVC SHEATHED. SINGLE CORE; GREY/BLUE 100MTR 35MM	1000m	29,699.70
6181YH PVC INSULATED/PVC SHEATHED. SINGLE CORE; GREY/BLUE 100MTR 6.0MM	1000m	5,010.90
6181YH PVC INSULATED/PVC SHEATHED. SINGLE CORE; GREY/BROWN 100MTR 25MM	1000m	19,509.30
6181YH PVC INSULATED/PVC SHEATHED. SINGLE CORE; GREY/BROWN 100MTR 35MM	1000m	29,699.70
6181YH PVC INSULATED/PVC SHEATHED. SINGLE CORE; GREY/BROWN 100MTR 6.0MM	1000m	5,010.90
6181YH PVC INSULATED/PVC SHEATHED. SINGLE CORE; GREY/BROWN 50MTR 25MM	1000m	20,971.20
6242BH 2 CORE FLAT XLPE INSULATED & LSZH SHEATHED C/W BARE CPC; 6242BH 2 CORE 1.5MM	1000m	1,720.80
6242BH 2 CORE FLAT XLPE INSULATED & LSZH SHEATHED C/W BARE CPC; 6242BH 2 CORE 10MM	1000m	18,303.70
6242BH 2 CORE FLAT XLPE INSULATED & LSZH SHEATHED C/W BARE CPC; 6242BH 2 CORE 16MM	1000m	26,692.70
6242BH 2 CORE FLAT XLPE INSULATED & LSZH SHEATHED C/W BARE CPC; 6242BH 2 CORE 2.5MM	1000m	2,378.40
6242BH 2 CORE FLAT XLPE INSULATED & LSZH SHEATHED C/W BARE CPC; 6242BH 2 CORE 4.0MM	1000m	7,162.20
6242BH 2 CORE FLAT XLPE INSULATED & LSZH SHEATHED C/W BARE CPC; 6242BH 2 CORE 6.0MM	1000m	10,205.60
6242YH PVC INSULATED/PVC SHEATHED. FLAT TWIN + EARTH; GREY 100MTR 1.5MM	1000m	1,753.60
6242YH PVC INSULATED/PVC SHEATHED. FLAT TWIN + EARTH; GREY 100MTR 10.0MM	1000m	15,992.50
6242YH PVC INSULATED/PVC SHEATHED. FLAT TWIN + EARTH; GREY 100MTR 16.0MM	1000m	25,489.60
6242YH PVC INSULATED/PVC SHEATHED. FLAT TWIN + EARTH; GREY 100MTR 2.5MM	1000m	2,401.50
6242YH PVC INSULATED/PVC SHEATHED. FLAT TWIN + EARTH; GREY 100MTR 4.0MM	1000m	8,161.80
6242YH PVC INSULATED/PVC SHEATHED. FLAT TWIN + EARTH; GREY 100MTR 6.0MM	1000m	9,694.10
6243BH 3 CORE FLAT XPLE INSULATED & LSZH SHEATHED C/W BARE CPC; 6243BH 3 CORE 1.0MM	1000m	4,791.50
6243BH 3 CORE FLAT XPLE INSULATED & LSZH SHEATHED C/W BARE CPC; 6243BH 3 CORE 1.5MM	1000m	6,086.00
6243YH PVC INSULATED/PVC SHEATHED. FLAT 3 CORE + EARTH; GREY 100MTR 1.5MM	1000m	5,833.30

BASIC PRICES : LABOUR, PLANT AND MATERIALS

V:ELECTRICAL SERVICES : WORKS OF ALTERATION/SMALL WORKS
CABLE & CABLE ACCESSORIES
GENERAL WIRING CABLES

Description	Unit	Price
6491B STRANDED LSZH SINGLE CORE; 6491B 1.5MM	1000m	427.10
6491B STRANDED LSZH SINGLE CORE; 6491B 10.0MM	1000m	2,598.20
6491B STRANDED LSZH SINGLE CORE; 6491B 16.0MM	1000m	3,673.80
6491B STRANDED LSZH SINGLE CORE; 6491B 2.5MM	1000m	583.90
6491B STRANDED LSZH SINGLE CORE; 6491B 25.0MM	1000m	7,318.30
6491B STRANDED LSZH SINGLE CORE; 6491B 35.0MM	1000m	9,459.70
6491B STRANDED LSZH SINGLE CORE; 6491B 4.0MM	1000m	1,028.00
6491B STRANDED LSZH SINGLE CORE; 6491B 50.0MM	1000m	10,867.50
6491B STRANDED LSZH SINGLE CORE; 6491B 6.0MM	1000m	1,528.60
6491X STRANDED PVC INSULATED. SINGLE CORE; BROWN 100MTR 1.5MM	1000m	731.60
6491X STRANDED PVC INSULATED. SINGLE CORE; BROWN 100MTR 10MM	1000m	6,496.20
6491X STRANDED PVC INSULATED. SINGLE CORE; BROWN 100MTR 16MM	1000m	10,255.10
6491X STRANDED PVC INSULATED. SINGLE CORE; BROWN 100MTR 2.5MM	1000m	1,078.10
6491X STRANDED PVC INSULATED. SINGLE CORE; BROWN 100MTR 25MM	1000m	10,619.70
6491X STRANDED PVC INSULATED. SINGLE CORE; BROWN 100MTR 35MM	1000m	13,212.20
6491X STRANDED PVC INSULATED. SINGLE CORE; BROWN 100MTR 4.0MM	1000m	2,390.20
6491X STRANDED PVC INSULATED. SINGLE CORE; BROWN 100MTR 50MM	1000m	19,269.10
6491X STRANDED PVC INSULATED. SINGLE CORE; BROWN 100MTR 6.0MM	1000m	3,474.40
PYRO MULTI-PLUS MULTI-PURPOSE CABLE SYSTEM - NEW RANGE; 4.0MM 4CORE + CPC 100MTR MP4HE4.0	1000m	7,614.00
PYRO MULTI-PLUS MULTI-PURPOSE CABLE SYSTEM - NEW RANGE; 6.0MM 2CORE + CPC 100MTR MP2HE6.0	1000m	8,920.00
PYRO MULTI-PLUS MULTI-PURPOSE CABLE SYSTEM - NEW RANGE; 6.0MM 3CORE + CPC 100MTR MP3HE6.0	1000m	10,384.00
PYRO MULTI-PLUS MULTI-PURPOSE CABLE SYSTEM - NEW RANGE; 6.0MM 4CORE + CPC 100MTR MP4HE6.0	1000m	12,304.00

INDUSTRIAL CABLES - STEEL WIRE ARMOURED

Description	Unit	Price
6942XL XLPE/PVC/SWA/PVC 2 CORE; XLPE/SWA STRANDED 25MM	m	16.47
6942XL XLPE/PVC/SWA/PVC 2 CORE; XLPE/SWA STRANDED 35MM	m	22.72
6942XL XLPE/PVC/SWA/PVC 2 CORE; XLPE/SWA STRANDED 1.5MM	1000m	3,800.90

BASIC PRICES : LABOUR, PLANT AND MATERIALS

V:ELECTRICAL SERVICES : WORKS OF ALTERATION/SMALL WORKS
CABLE & CABLE ACCESSORIES

INDUSTRIAL CABLES - STEEL WIRE ARMOURED

Description	Unit	Price
6942XL XLPE/PVC/SWA/PVC 2 CORE; XLPE/SWA STRANDED 10MM	1000m	13,098.50
6942XL XLPE/PVC/SWA/PVC 2 CORE; XLPE/SWA STRANDED 16MM	1000m	16,038.40
6942XL XLPE/PVC/SWA/PVC 2 CORE; XLPE/SWA STRANDED 2.5MM	1000m	4,372.50
6942XL XLPE/PVC/SWA/PVC 2 CORE; XLPE/SWA STRANDED 4.0MM	1000m	6,905.80
6942XL XLPE/PVC/SWA/PVC 2 CORE; XLPE/SWA STRANDED 6.0MM	1000m	8,720.10
6943XL XLPE/PVC/SWA/PVC 3 CORE; XLPE/SWA STRANDED 1.5MM	1000m	4,265.30
6943XL XLPE/PVC/SWA/PVC 3 CORE; XLPE/SWA STRANDED 10MM	1000m	18,358.60
6943XL XLPE/PVC/SWA/PVC 3 CORE; XLPE/SWA STRANDED 16MM	1000m	22,705.40
6943XL XLPE/PVC/SWA/PVC 3 CORE; XLPE/SWA STRANDED 2.5MM	1000m	5,232.80
6943XL XLPE/PVC/SWA/PVC 3 CORE; XLPE/SWA STRANDED 25MM	1000m	19.59
6943XL XLPE/PVC/SWA/PVC 3 CORE; XLPE/SWA STRANDED 35MM	1000m	26.52
6943XL XLPE/PVC/SWA/PVC 3 CORE; XLPE/SWA STRANDED 4.0MM	1000m	8,109.40
6943XL XLPE/PVC/SWA/PVC 3 CORE; XLPE/SWA STRANDED 6.0MM	1000m	10,284.90
6944XL XLPE/PVC/SWA/PVC 4 CORE; XLPE/SWA STRANDED 1.5MM	1000m	4,739.50
6944XL XLPE/PVC/SWA/PVC 4 CORE; XLPE/SWA STRANDED 10MM	1000m	21,403.50
6944XL XLPE/PVC/SWA/PVC 4 CORE; XLPE/SWA STRANDED 16MM	1000m	25,171.20
6944XL XLPE/PVC/SWA/PVC 4 CORE; XLPE/SWA STRANDED 2.5MM	1000m	6,047.90
6944XL XLPE/PVC/SWA/PVC 4 CORE; XLPE/SWA STRANDED 25MM	1000m	26.76
6944XL XLPE/PVC/SWA/PVC 4 CORE; XLPE/SWA STRANDED 35MM	1000m	35.67
6944XL XLPE/PVC/SWA/PVC 4 CORE; XLPE/SWA STRANDED 4.0MM	1000m	9,665.30
6944XL XLPE/PVC/SWA/PVC 4 CORE; XLPE/SWA STRANDED 6.0MM	1000m	13,170.70

PYROTENAX MINERAL INSULATED CABLES AND ACCESSORIES

Description	Unit	Price
COMPRESSION GLANDS - BRASS (HEAVY DUTY); COMPRESSION GLAND FOR 2H2.5 CABLE RGM2H2.5	Item	32.42
COMPRESSION GLANDS - BRASS (HEAVY DUTY); COMPRESSION GLAND FOR 2H4 CABLE RGM2H4	Item	32.42

BASIC PRICES : LABOUR, PLANT AND MATERIALS

V:ELECTRICAL SERVICES : WORKS OF ALTERATION/SMALL WORKS
CABLE & CABLE ACCESSORIES
PYROTENAX MINERAL INSULATED CABLES AND ACCESSORIES

Description	Unit	Price
COMPRESSION GLANDS - BRASS (HEAVY DUTY); COMPRESSION GLAND FOR 2H6 CABLE RGM2H6	Item	32.42
COMPRESSION GLANDS - BRASS (HEAVY DUTY); COMPRESSION GLAND FOR 3H1.5 CABLE RGM3H1.5	Item	32.42
COMPRESSION GLANDS - BRASS (HEAVY DUTY); COMPRESSION GLAND FOR 3H2.5 CABLE RGM3H2.5	Item	32.42
COMPRESSION GLANDS - BRASS (HEAVY DUTY); COMPRESSION GLAND FOR 3H4 CABLE RGM3H4	Item	32.42
COMPRESSION GLANDS - BRASS (HEAVY DUTY); COMPRESSION GLAND FOR 4H1.5 CABLE RGM4H1.5	Item	32.42
COMPRESSION GLANDS - BRASS (HEAVY DUTY); COMPRESSION GLAND FOR 4H4 CABLE RGM4H4	Item	21.60
COMPRESSION GLANDS - BRASS (HEAVY DUTY); COMPRESSION GLAND FOR 4H4 CABLE RGM4H4	Item	21.60
COMPRESSION GLANDS - BRASS (HEAVY DUTY); COMPRESSION GLAND FOR 4H6 CABLE RGM4H6	Item	21.60
COMPRESSION GLANDS - BRASS (LIGHT DUTY); COMPRESSION GLAND FOR 2L1.5 CABLE RGM2L1.5	Item	6.36
COMPRESSION GLANDS - BRASS (LIGHT DUTY); COMPRESSION GLAND FOR 2L2.5 CABLE RGM2L2.5	Item	6.36
COMPRESSION GLANDS - BRASS (LIGHT DUTY); COMPRESSION GLAND FOR 3L1.5 CABLE RGM3L1.5	Item	6.36
COMPRESSION GLANDS - BRASS (LIGHT DUTY); COMPRESSION GLAND FOR 3L2.5 CABLE RGM3L2.5	Item	6.36
COMPRESSION GLANDS - BRASS (LIGHT DUTY); COMPRESSION GLAND FOR 4L1.5 CABLE RGM4L1.5	Item	6.36
COMPRESSION GLANDS - BRASS (LIGHT DUTY); COMPRESSION GLAND FOR 4L2.5 CABLE RGM4L2.5	Item	6.36
GLAND SHROUDS - PVC AND LSF TYPES; RED GLAND SHROUD LSF 25MM RHGMM25	Item	19.63
GLAND SHROUDS - PVC AND LSF TYPES; RED GLAND SHROUD PVC 20MM RHG20	Item	10.90
GLAND SHROUDS - PVC AND LSF TYPES; RED LSF 25MM RHGMM25	Item	19.63
GLAND SHROUDS - PVC AND LSF TYPES; RED PVC 20MM RHG20	Item	10.90
GLAND SHROUDS - PVC AND LSF TYPES; WHITE GLAND SHROUD PVC 20MM RHG20	Item	10.90
GLAND SHROUDS - PVC AND LSF TYPES; WHITE PVC 20MM RHG20	Item	10.90
HEAVY DUTY - LSF SHEATHED ORANGE - 750V GRADE - (CUT TO LENGTH SERVICE); 4.0MM 3CORE CCM3H4	1000m	2,361.00
LIGHT DUTY - BARE COPPER SHEATHED - 500V GRADE - 100MTR DRUM; 1.5MM 2CORE CC2L1.5	100m	693.30
LIGHT DUTY - BARE COPPER SHEATHED - 500V GRADE - 100MTR DRUM; 1.5MM 2CORE CC2L1.5	100m	693.30

© NSR 2008-2009

BASIC PRICES : LABOUR, PLANT AND MATERIALS

V:ELECTRICAL SERVICES : WORKS OF ALTERATION/SMALL WORKS
CABLE & CABLE ACCESSORIES

PYROTENAX MINERAL INSULATED CABLES AND ACCESSORIES

Description	Unit	Price
LIGHT DUTY - BARE COPPER SHEATHED - 500V GRADE - 100MTR DRUM; 1.5MM 3CORE CC3L1.5	100m	917.20
LIGHT DUTY - BARE COPPER SHEATHED - 500V GRADE - 100MTR DRUM; 1.5MM 4CORE CC4L1.5	100m	1,077.40
LIGHT DUTY - BARE COPPER SHEATHED - 500V GRADE - 100MTR DRUM; 1.5MM 7CORE CC7L1.5	100m	1,664.10
LIGHT DUTY - BARE COPPER SHEATHED - 500V GRADE - 100MTR DRUM; 2.5MM 2CORE CC2L2.5	100m	900.60
LIGHT DUTY - LSF SHEATHED BLACK - 500 GRADE - 100MTR DRUM; 1.5MM 4CORE CCM4L1.5	100m	1,202.50
LIGHT DUTY - LSF SHEATHED ORANGE - 500V GRADE - (CUT TO LENGTH SERVICE); 2.5MM 3CORE CCM3L2.5	m	15.20
LIGHT DUTY - LSF SHEATHED ORANGE - 500V GRADE - (CUT TO LENGTH SERVICE); 2.5MM 4CORE CCM4L2.5	m	18.45
LIGHT DUTY - LSF SHEATHED ORANGE - 500V GRADE - (CUT TO LENGTH SERVICE); 4.0MM 2CORE CCM2L4	m	14.47
LIGHT DUTY - LSF SHEATHED ORANGE - 500V GRADE - 100MTR DRUM; 1.0MM 2CORE CCM2L1	100m	686.20
LIGHT DUTY - LSF SHEATHED ORANGE - 500V GRADE - 100MTR DRUM; 1.0MM 3CORE CCM3L1	100m	832.50
LIGHT DUTY - LSF SHEATHED ORANGE - 500V GRADE - 100MTR DRUM; 1.5MM 2CORE CCM2L1.5	100m	761.60
LIGHT DUTY - LSF SHEATHED ORANGE - 500V GRADE - 100MTR DRUM; 1.5MM 2CORE 100MTR 500V LSF 2L1.5	100m	761.60
LIGHT DUTY - LSF SHEATHED ORANGE - 500V GRADE - 100MTR DRUM; 1.5MM 3CORE 500V LSF 3L1.5	100m	1,015.90
LIGHT DUTY - LSF SHEATHED ORANGE - 500V GRADE - 100MTR DRUM; 1.5MM 3CORE CCM3L1.5	100m	1,015.90
LIGHT DUTY - LSF SHEATHED ORANGE - 500V GRADE - 100MTR DRUM; 1.5MM 4CORE CCM4L1.5	100m	1,202.50
LIGHT DUTY - LSF SHEATHED ORANGE - 500V GRADE - 100MTR DRUM; 2.5MM 2CORE CCM2L2.5	m	9.62
LIGHT DUTY - LSF SHEATHED ORANGE - 500V GRADE - 100MTR DRUM; 2.5MM 3CORE CCM3L2.5	100m	1,520.30
LIGHT DUTY - LSF SHEATHED RED - 500V GRADE - (CUT TO LENGTH SERVICE); CUTS 4.0MM 2CORE CCM2L4	m	14.47
LIGHT DUTY - LSF SHEATHED RED - 500V GRADE - 100MTR DRUM; 1.0MM 4CORE 100MTR CCM4L1	100m	978.40
LIGHT DUTY - LSF SHEATHED WHITE - 500V GRADE - (CUT TO LENGTH SERVICE); CUTS 2.5MM 4CORE CCM4L2.5	m	18.15
SEAL ASSEMBLIES - (HEAVY DUTY) EARTH TAIL TYPE; 2H6 CABLE RPSL2H6	Item	27.90
SEAL ASSEMBLIES - (HEAVY DUTY) STANDARD SCREW-ON TYPE; SEAL SCREW-ON FOR 2H4 CABLE RPS2H4	Item	21.15

BASIC PRICES : LABOUR, PLANT AND MATERIALS

V:ELECTRICAL SERVICES : WORKS OF ALTERATION/SMALL WORKS
CABLE & CABLE ACCESSORIES

PYROTENAX MINERAL INSULATED CABLES AND ACCESSORIES

Description	Unit	Price
SEAL ASSEMBLIES - (HEAVY DUTY) STANDARD SCREW-ON TYPE; SEAL SCREW-ON FOR 2H4 CABLE RPS2H4	Item	21.15
SEAL ASSEMBLIES - (HEAVY DUTY) STANDARD SCREW-ON TYPE; SEAL SCREW-ON FOR 2H6 CABLE RPS2H6	Item	21.15
SEAL ASSEMBLIES - (HEAVY DUTY) STANDARD SCREW-ON TYPE; SEAL SCREW-ON FOR 3H1.5 CABLE RPS3H1.5	Item	21.54
SEAL ASSEMBLIES - (HEAVY DUTY) STANDARD SCREW-ON TYPE; SEAL SCREW-ON FOR 3H2.5 CABLE RPS3H2.5	Item	21.54
SEAL ASSEMBLIES - (HEAVY DUTY) STANDARD SCREW-ON TYPE; SEAL SCREW-ON FOR 3H4 CABLE RPS3H4	Item	21.54
SEAL ASSEMBLIES - (HEAVY DUTY) STANDARD SCREW-ON TYPE; SEAL SCREW-ON FOR 4H1.5 CABLE RPS4H1.5	Item	21.54
SEAL ASSEMBLIES - (HEAVY DUTY) STANDARD SCREW-ON TYPE; SEAL SCREW-ON FOR 4H2.5 CABLE RPS4H2.5	Item	21.54
SEAL ASSEMBLIES - (HEAVY DUTY) STANDARD SCREW-ON TYPE; SEAL SCREW-ON FOR 4H4 CABLE RPS4H4	Item	19.97
SEAL ASSEMBLIES - (HEAVY DUTY) STANDARD SCREW-ON TYPE; SEAL SCREW-ON FOR 4H6 CABLE RPS4H6	Item	19.97
SEAL ASSEMBLIES - (LIGHT DUTY) EARTH TAIL TYPE; EARTH TAIL SEAL FOR 2L1.5 CABLE RPSL2L1.5	Item	39.01
SEAL ASSEMBLIES - (LIGHT DUTY) EARTH TAIL TYPE; EARTH TAIL SEAL FOR 3L2.5 CABLE RPSL3L2.5	Item	40.46
SEAL ASSEMBLIES - (LIGHT DUTY) STANDARD SCREW-ON TYPE; SEAL SCREW-ON FOR 2L1.5 CABLE RPS2L1.5	Item	20.79
SEAL ASSEMBLIES - (LIGHT DUTY) STANDARD SCREW-ON TYPE; SEAL SCREW-ON FOR 2L2.5 CABLE RPS2L2.5	Item	20.79
SEAL ASSEMBLIES - (LIGHT DUTY) STANDARD SCREW-ON TYPE; SEAL SCREW-ON FOR 3L1.5 CABLE RPS3L1.5	Item	21.54
SEAL ASSEMBLIES - (LIGHT DUTY) STANDARD SCREW-ON TYPE; SEAL SCREW-ON FOR 3L2.5 CABLE RPS3L2.5	Item	21.54
SEAL ASSEMBLIES - (LIGHT DUTY) STANDARD SCREW-ON TYPE; SEAL SCREW-ON FOR 4L1.5 CABLE RPS4L1.5	Item	21.54
SEAL ASSEMBLIES - (LIGHT DUTY) STANDARD SCREW-ON TYPE; SEAL SCREW-ON FOR 4L2.5 CABLE RPS4L2.5	Item	21.54

BASIC PRICES : LABOUR, PLANT AND MATERIALS

V:ELECTRICAL SERVICES : WORKS OF ALTERATION/SMALL WORKS
CABLE MANAGEMENT SYSTEMS
ADAPTABLE BOXES

Description	Unit	Price
APPLEBY (STEEL) ADAPTABLE BOX; KO 100X100X100 BLACK AB559PB	Item	7.47
APPLEBY (STEEL) ADAPTABLE BOX; KO 100X100X100 GALVANISED AB559KG	Item	12.35
APPLEBY (STEEL) ADAPTABLE BOX; KO 100X100X50 GALVANISED AB557KG	Item	5.36
APPLEBY (STEEL) ADAPTABLE BOX; KO 100X100X75 BLACK AB558KB	Item	4.03
APPLEBY (STEEL) ADAPTABLE BOX; KO 100X100X75 GALVANISED AB558KG	Item	6.97
APPLEBY (STEEL) ADAPTABLE BOX; KO 150X100X75 BLACK AB564KB	Item	4.84
APPLEBY (STEEL) ADAPTABLE BOX; KO 225X225X100 BLACK AB580B	Item	11.32
APPLEBY (STEEL) ADAPTABLE BOX; KO 225X225X75 BLACK AB579KB	Item	9.96
APPLEBY (STEEL) ADAPTABLE BOX; KO 300X150X75 BLACK AB582KB	Item	10.08
APPLEBY (STEEL) ADAPTABLE BOX; KO 300X300X100 BLACK AB590KB	Item	26.90
APPLEBY (STEEL) ADAPTABLE BOX; KO 300X300X75 BLACK AB589KB	Item	13.74
APPLEBY (STEEL) ADAPTABLE BOX; KO 75X75X75 BLACK AB553KB	Item	3.60
APPLEBY (STEEL) ADAPTABLE BOX; KO 75X75X75 GALVANISED AB553KG	Item	5.86
APPLEBY (STEEL) ADAPTABLE BOX; PLAIN 150X150X100 BLACK AB569PB	Item	8.36
APPLEBY (STEEL) ADAPTABLE BOX; PLAIN 150X150X100 GALVANISED AB569PG	Item	13.90
APPLEBY (STEEL) ADAPTABLE BOX; PLAIN 150X150X50 GALVANISED AB566PG	Item	7.64
APPLEBY (STEEL) ADAPTABLE BOX; PLAIN 150X150X75 GALVANISED AB567PG	Item	9.33
APPLEBY (STEEL) ADAPTABLE BOX; PLAIN 150X75X50 BLACK AB561PB	Item	6.01
APPLEBY (STEEL) ADAPTABLE BOX; PLAIN 225X225X100 GALVANISED AB580PG	Item	18.82
APPLEBY (STEEL) ADAPTABLE BOX; PLAIN 225X225X75 GALVANISED AB579PG	Item	16.41
APPLEBY (STEEL) ADAPTABLE BOX; PLAIN 300X300X100 GALVANISED AB590PG	Item	36.64
APPLEBY (STEEL) ADAPTABLE BOX; PLAIN 300X300X150 GALVANISED AB591KG	Item	43.96
APPLEBY (STEEL) ADAPTABLE BOX; PLAIN 300X300X75 GALVANISED AB589PG	Item	22.87

BASIC PRICES : LABOUR, PLANT AND MATERIALS

V:ELECTRICAL SERVICES : WORKS OF ALTERATION/SMALL WORKS
CABLE MANAGEMENT SYSTEMS
CABLE TRAY (STEEL) AND FITTINGS

Description	Unit	Price
MITA AEMSA ACCESSORIES; COUPLER 250MM EJOINTE	Item	2.88
MITA AEMSA WIRE CABLE TRAY; 70X500MM WIRE CABLE TRAY 3MTR LENGTH REZ70X500	Item	59.27
MITA AEMSA WIRE CABLE TRAY; 70X600MM WIRE CABLE TRAY 3MTR LENGTH REZ70X600	Item	79.50
NEWLEC CABLE TRAY - LIGHT DUTY - (PRE-GALVANISED); 100MM WIDE 3MTR LENGTH NLZL100SL	Item	9.95
NEWLEC CABLE TRAY - LIGHT DUTY - (PRE-GALVANISED); 150MM WIDE 3MTR LENGTH NLZL150SL	Item	12.95
NEWLEC CABLE TRAY - LIGHT DUTY - (PRE-GALVANISED); 225MM WIDE 3MTR LENGTH NLZL225SL	Item	22.58
NEWLEC CABLE TRAY - LIGHT DUTY - (PRE-GALVANISED); 50MM WIDE 3MTR LENGTH NLZL50SL	Item	6.25
NEWLEC CABLE TRAY - LIGHT DUTY - (PRE-GALVANISED); 75MM WIDE 3MTR LENGTH NLZL75SL	Item	7.50
NEWLEC CABLE TRAY - MEDIUM DUTY - (PRE-GALVANISED); 150MM WIDE 3MTR LENGTH NLZM150SL	Item	13.40
NEWLEC CABLE TRAY - MEDIUM DUTY - (PRE-GALVANISED); 225MM WIDE 3MTR LENGTH NLZM225SL	Item	17.95
NEWLEC CABLE TRAY - MEDIUM DUTY - (PRE-GALVANISED); 300MM WIDE 3MTR LENGTH NLZM300SL	Item	30.26
NEWLEC COUPLERS; COUPLER (PAIR) NLZMC	Item	3.25
PGI COUPLER - MRF - GALVANISED; COUPLER ASSEMBLY (PACK 2) MRC/G	Item	8.46
PGI COUPLER - SRF - GALVANISED; ASSEMBLY PACK2 HRC/G	PAIR	11.78
PGI TRAY - MRF [MEDIUM RETURN FLANGE] GALVANISED; 3MTR 100MM WIDTH MRL/100/G/3M	Item	45.79
PGI TRAY - MRF [MEDIUM RETURN FLANGE] GALVANISED; 3MTR 150MM WIDTH MRL/150/G/3M	Item	58.24
PGI TRAY - MRF [MEDIUM RETURN FLANGE] GALVANISED; 3MTR 225MM WIDTH MRL/225/G/3M	Item	76.49
PGI TRAY - MRF [MEDIUM RETURN FLANGE] GALVANISED; 3MTR 300MM WIDTH MRL/300/G/3M	Item	129.42
PGI TRAY - SRF - [HEAVY RETURN FLANGE] GALVANISED; 3MTR 100MM WIDTH HRL/100/G/3M	Item	80.30
PGI TRAY - SRF - [HEAVY RETURN FLANGE] GALVANISED; 3MTR 150MM WIDTH HRL/150/G/3M	Item	91.25
PGI TRAY - SRF - [HEAVY RETURN FLANGE] GALVANISED; 3MTR 225MM WIDTH HRL/225/G/3M	Item	123.78

BASIC PRICES : LABOUR, PLANT AND MATERIALS

V:ELECTRICAL SERVICES : WORKS OF ALTERATION/SMALL WORKS
CABLE MANAGEMENT SYSTEMS
CABLE TRAY (STEEL) AND FITTINGS

Description	Unit	Price
PGI TRAY - SRF - [HEAVY RETURN FLANGE] GALVANISED; 3MTR 300MM WIDTH HRL/300/G/3M	Item	142.64
PGI TRAY - SRF - [HEAVY RETURN FLANGE] GALVANISED; 3MTR 450MM WIDTH HRL/450/G/3M	Item	259.77
PGI TRAY - SRF - [HEAVY RETURN FLANGE] GALVANISED; 3MTR 600MM WIDTH HRL/600/G/3M	Item	365.05

CABLE TRUNKING (STEEL) AND FITTINGS

Description	Unit	Price
BARTON 2 AND 3 COMPARTMENT TRUNKING - GALVANISED - BS4678; 100X100MM 2 COMPARTMENT 3MTR LENGTH C/W LID + CONNECTOR QF442	Item	73.30
BARTON 2 AND 3 COMPARTMENT TRUNKING - GALVANISED - BS4678; 100X50MM 2 COMPARTMENT 3MTR LENGTH C/W LID + CONNECTOR QF422	Item	58.33
BARTON CABLE TRUNKING - GALVANISED - BS4678; 100X100MM C/W LID + CONNECTOR 3MTR LENGTH QF44	Item	59.79
BARTON CABLE TRUNKING - GALVANISED - BS4678; 100X50MM C/W LID + CONNECTOR 3MTR LENGTH QF42	Item	49.83
BARTON CABLE TRUNKING - GALVANISED - BS4678; 150X100MM C/W LID + CONNECTOR 3MTR LENGTH QF64	Item	82.79
BARTON CABLE TRUNKING - GALVANISED - BS4678; 150X150MM C/W LID + CONNECTOR 3MTR LENGTH QF66	Item	103.53
BARTON CABLE TRUNKING - GALVANISED - BS4678; 50X50MM C/W LID + CONNECTOR 3MTR LENGTH QF22	Item	32.27
BARTON CABLE TRUNKING - GALVANISED - BS4678; 75X50MM C/W LID + CONNECTOR 3MTR LENGTH QF32	Item	42.37
BARTON CABLE TRUNKING - GALVANISED - BS4678; 75X75MM C/W LID + CONNECTOR 3MTR LENGTH QF33	Item	46.37

CHANNEL (STEEL AND PLASTIC)

Description	Unit	Price
CHANNEL (STEEL GALVANISED); STEEL GALVANISED 2MTR 12MM	Item	0.55
CHANNEL (STEEL GALVANISED); STEEL GALVANISED 2MTR 25MM	Item	0.65
CHANNEL (STEEL GALVANISED); STEEL GALVANISED 2MTR 38MM	Item	0.75
CHANNEL (STEEL GALVANISED); STEEL GALVANISED 2MTR 50MM	Item	1.25
MK (PVC) CHANNEL; WHITE 12.5MM 2MTR LENGTH REC1	Item	2.60
MK (PVC) CHANNEL; WHITE 25MM 2MTR LENGTH REC2	Item	3.56
MK (PVC) CHANNEL; WHITE 38MM 2MTR LENGTH REC3	Item	5.24

CONDUIT FITTINGS MALLEABLE

Description	Unit	Price
BARTON BOXES - ANGLE; B/E SMALL 25MM M183	Item	7.54
BARTON BOXES - ANGLE; B/E SMALL 20MM M183	Item	5.82

BASIC PRICES : LABOUR, PLANT AND MATERIALS

V:ELECTRICAL SERVICES : WORKS OF ALTERATION/SMALL WORKS
CABLE MANAGEMENT SYSTEMS
CONDUIT FITTINGS MALLEABLE

Description	Unit	Price
BARTON BOXES - BRANCH 3 WAY (SMALL Y); B/E SMALL Y 25MM M196 Y	Item	18.61
BARTON BOXES - BRANCH 3 WAY (SMALL Y); B/E SMALL Y 20MM M196 Y	Item	11.22
BARTON BOXES - BRANCH 3 WAY (SMALL Y); GALV. SMALL Y 20MM M196Y	Item	12.04
BARTON BOXES - BRANCH 3 WAY (SMALL Y); GALV. SMALL Y 25MM M196 Y	Item	18.86
BARTON BOXES - BRANCH 4 WAY (SMALL X); B/E SMALL 20MM M185	Item	7.57
BARTON BOXES - BRANCH 4 WAY (SMALL X); B/E SMALL 25MM M185	Item	11.11
BARTON BOXES - BRANCH 4 WAY (SMALL X); GALV. SMALL 20MM M185	Item	7.84
BARTON BOXES - BRANCH 4 WAY (SMALL X); GALV. SMALL 25MM M185	Item	11.26
BARTON BOXES - TANGENT; GALV. SMALL TANGENT ANGLE 25MM M192	Item	16.92
BARTON BOXES - TANGENT; GALV. SMALL TANGENT THROUGH 20MM M192	Item	12.54
BARTON BOXES - TERMINAL; B/E SMALL 20MM M181	Item	4.56
BARTON BOXES - TERMINAL; B/E SMALL 25MM M181	Item	6.32
BARTON BOXES - TERMINAL; GALV. SMALL 20MM M181	Item	4.89
BARTON BOXES - TERMINAL; GALV. SMALL 25MM M181	Item	6.78
BARTON BOXES - THROUGH; B/E SMALL 20MM M182	Item	5.42
BARTON BOXES - THROUGH; B/E SMALL 25MM M182	Item	7.54
BARTON BOXES - THROUGH; GALV. SMALL 20MM M182	Item	5.82
BARTON BOXES - THROUGH; GALV. SMALL 25MM M182	Item	8.09
CONDUIT BOX LIDS - PLAIN; LIGHT PATTERN SMALL BLACK R281	Item	0.67
CONDUIT BOX LIDS - PLAIN; LIGHT PATTERN SMALL GALVANISED R281	Item	0.56

CONDUIT FITTINGS PLASTIC

Description	Unit	Price
MARSHALL-TUFFLEX CIRCULAR BOXES; 20MM BLACK 4WAY 2MRB6	Item	4.80
MARSHALL-TUFFLEX CIRCULAR BOXES; 20MM BLACK ANGLE 2MRB4	Item	3.76
MARSHALL-TUFFLEX CIRCULAR BOXES; 20MM BLACK TERMINAL 2MRB2	Item	3.37
MARSHALL-TUFFLEX CIRCULAR BOXES; 20MM BLACK THROUGH 2MRB3	Item	3.76
MARSHALL-TUFFLEX CIRCULAR BOXES; 20MM BLACK Y (BRANCH 3 WAY) 2MRB14	Item	8.79

BASIC PRICES : LABOUR, PLANT AND MATERIALS

V:ELECTRICAL SERVICES : WORKS OF ALTERATION/SMALL WORKS
CABLE MANAGEMENT SYSTEMS
CONDUIT FITTINGS PLASTIC

	MARSHALL-TUFFLEX CIRCULAR BOXES; 25MM BLACK 4WAY 3MRB6	Item	7.22
	MARSHALL-TUFFLEX CIRCULAR BOXES; 25MM BLACK ANGLE 3MRB4	Item	5.78
	MARSHALL-TUFFLEX CIRCULAR BOXES; 25MM BLACK TERMINAL 3MRB2	Item	5.33
	MARSHALL-TUFFLEX CIRCULAR BOXES; 25MM BLACK THROUGH 3MRB3	Item	5.78
	MARSHALL-TUFFLEX CIRCULAR BOXES; 25MM WHITE Y (BRANCH 3 WAY) 3MRB14	Item	10.97
	MARSHALL-TUFFLEX CIRCULAR LID; 50MM DIA. BLACK MCL1	Item	0.97

CONDUIT PLASTIC

	MK LIGHT GAUGE CONDUIT; WHITE 16MM 3MTR HLG1	Item	6.18
	MK LIGHT GAUGE CONDUIT; WHITE 20MM 3MTR HLG2	Item	6.75
	MK LIGHT GAUGE CONDUIT; WHITE 25MM 3MTR HLG3	Item	11.25

CONDUIT STEEL

	CONDUIT BLACK ENAMEL; HGSW (3.75MTR LENGTH) 20MM	Item	14.89
	CONDUIT BLACK ENAMEL; HGSW (3.75MTR LENGTH) 25MM	Item	20.31
	CONDUIT BLACK ENAMEL; HGSW (3.75MTR LENGTH) 32MM	Item	32.00
	CONDUIT GALVANISED; HGSW (3.75MTR LENGTH) 20MM HDG	Item	19.88
	CONDUIT GALVANISED; HGSW (3.75MTR LENGTH) 25MM HDG	Item	24.68
	CONDUIT GALVANISED; HGSW (3.75MTR LENGTH) 32MM HDG	Item	38.18

FLEXIBLE CONDUIT (METALLIC) AND FITTINGS

	ADAPTAFLEX BRASS ADAPTORS - FOR SP TYPE CONDUIT; ADAPTOR 25MM TYPE-A SP25/M25A	Item	5.87
	ADAPTAFLEX BRASS ADAPTORS - FOR SP TYPE CONDUIT; ADAPTOR 32MM TYPE-A SP32/M32A	Item	10.86
	ADAPTAFLEX CONDUIT ACCESSORIES; LOCKNUT BRASS 20MM LNB/M20	Item	0.49
	ADAPTAFLEX CONDUIT ACCESSORIES; LOCKNUT BRASS 25MM LNB/M25	Item	0.79
	ADAPTAFLEX CONDUIT ACCESSORIES; LOCKNUT BRASS 32MM LNB/M32	Item	1.11
	ADAPTAFLEX FLEXIBLE CONDUIT STEEL/PVC; 25MTR STEEL/PVC COVERED 20MM SP20	25m	110.87
	ADAPTAFLEX FLEXIBLE CONDUIT STEEL/PVC; 25MTR STEEL/PVC COVERED 25MM SP25	25m	151.42

© NSR 2008-2009

BASIC PRICES : LABOUR, PLANT AND MATERIALS

V:ELECTRICAL SERVICES : WORKS OF ALTERATION/SMALL WORKS

CABLE MANAGEMENT SYSTEMS

FLEXIBLE CONDUIT (METALLIC) AND FITTINGS

Description	Unit	Price
ADAPTAFLEX FLEXIBLE CONDUIT STEEL/PVC; 25MTR STEEL/PVC COVERED 32MM SP32	25m	235.69
NEWLEC STEEL/PVC FLEXIBLE CONDUIT AND FITTINGS - NEW RANGE; 20MM CONDUIT KIT (KIT OF 10MTR CONDUIT + 10 CONNECTORS) NLSP20KIT	Item	42.15

TRUNKING AND FITTINGS PLASTIC

Description	Unit	Price
MARSHALL-TUFFLEX MINI TRUNKING - WHITE; 16X16MM 3MTR LENGTH MMT1	Item	12.76
MARSHALL-TUFFLEX MINI TRUNKING - WHITE; 16X25MM 3MTR LENGTH MMT2	Item	15.96
MARSHALL-TUFFLEX MINI TRUNKING - WHITE; 16X38MM 3MTR LENGTH MMT3	Item	20.28
MARSHALL-TUFFLEX MINI TRUNKING - WHITE; 25X38MM 3MTR LENGTH MMT4	Item	24.43
MARSHALL-TUFFLEX MINI TRUNKING - WHITE; TWIN COMPARTMENT 25X38MM 3MTR LENGTH MMT4C	Item	28.72
MITA TRK HEAVY DUTY TRUNKING - WHITE; 150X150MM 3MTR LENGTH TRK150W	Item	201.72

DISTRIBUTION EQUIPMENT

CONSUMER UNITS 250V + MCB S

Description	Unit	Price
MEM CONSUMER UNITS MEMERA 2000AD; 12WAY C/W 100A 30MA RCD INSULATED AD12HE	Item	98.99
MEM CONSUMER UNITS MEMERA 2000AD; 4WAY C/W 100A 30MA RCD INSULATED AD4HE	Item	72.63
MEM CONSUMER UNITS MEMERA 2000AD; 6WAY C/W 100A 30MA RCD INSULATED AD6HE	Item	78.93
MEM CONSUMER UNITS MEMERA 2000AD; 8WAY C/W 100A 30MA RCD INSULATED AD8HE	Item	82.50
NEWLEC INSULATED CONSUMER UNITS; CONSUMER UNIT 2WAY + 63A MAIN SWITCH INCOMER NLCU2PSW	Item	14.95
NEWLEC INSULATED CONSUMER UNITS; CONSUMER UNIT 5WAY + 100A MAIN SWITCH INCOMER NLCU5PSW	Item	17.95
NEWLEC INSULATED CONSUMER UNITS; CONSUMER UNIT 8WAY + 100A MAIN SWITCH INCOMER NLCU8PSW	Item	19.95
WYLEX CONSUMER UNITS - NH RANGE (SPLITLOAD WITH TIME DELAY) - NEW RANGE; 12WAY SPLITLOAD CONSUMER UNIT 6WAY 100A 100MA 6WAY 80A 30MA - INSULATED NHSTM6604	Item	193.79
WYLEX CONSUMER UNITS - NH RANGE (SPLITLOAD WITH TIME DELAY) - NEW RANGE; 12WAY SPLITLOAD CONSUMER UNIT 7WAY 100A 100MA 5WAY 80A 30MA - INSULATED NHSTM5704	Item	193.79
WYLEX CONSUMER UNITS - NH RANGE (SPLITLOAD WITH TIME DELAY) - NEW RANGE; 9WAY SPLITLOAD CONSUMER UNIT 5WAY 100A 100MA 4WAY 80A 30MA - INSULATED NHSTM4504	Item	185.78

BASIC PRICES : LABOUR, PLANT AND MATERIALS

V:ELECTRICAL SERVICES : WORKS OF ALTERATION/SMALL WORKS
DISTRIBUTION EQUIPMENT

DISTRIBUTION BOARDS - (MCB)

Description	Unit	Price
CRABTREE C50 TRIPLE POLE MCBS; 30A TRIPLE POLE 53/30	Item	75.88
CRABTREE C50 TRIPLE POLE MCBS; 40A TRIPLE POLE 53/40	Item	79.47
CRABTREE C50 TRIPLE POLE MCBS; 50A TRIPLE POLE 53/50	Item	79.47
CRABTREE DISTRIBUTION BOARDS POLESTAR - TP&N; TPN 10WAY WITH 125A TP MAIN SWITCH INCOMING DEVICE 1810/OBF	Item	198.66
CRABTREE DISTRIBUTION BOARDS POLESTAR - TP&N; TPN 12WAY TYPE B 1812/OB	Item	190.93
CRABTREE DISTRIBUTION BOARDS POLESTAR - TP&N; TPN 4WAY TYPE B 1804/OB	Item	116.45
CRABTREE DISTRIBUTION BOARDS POLESTAR - TP&N; TPN 4WAY WITH 125A TP MAIN SWITCH INCOMING DEVICE 1804/OBF	Item	133.19
CRABTREE DISTRIBUTION BOARDS POLESTAR - TP&N; TPN 6WAY TYPE B + 125A TP MAIN SWITCH INCOMING DEVICE 1806/OBF	Item	154.46
CRABTREE DISTRIBUTION BOARDS POLESTAR - TP&N; TPN 8WAY TYPE B 1808/OB	Item	152.28
CRABTREE DISTRIBUTION BOARDS POLESTAR - TP&N; TPN 8WAY WITH 125A TP MAIN SWITCH INCOMING DEVICE 1808/OBF	Item	173.40
CRABTREE POLESTAR DOUBLE POLE MCBS; 20A DOUBLE POLE TYPE-C 62C20	Item	27.63
CRABTREE POLESTAR DOUBLE POLE MCBS; 40A DOUBLE POLE TYPE-C 62C40	Item	29.19
CRABTREE POLESTAR DOUBLE POLE MCBS; 63A DOUBLE POLE TYPE-C 62C63	Item	29.19
CRABTREE POLESTAR MCB/RCDS; 6A MCB/RCD TYPE-C 602C063	Item	68.46
CRABTREE POLESTAR RCDS; 100A 30MA RCD TP/N 641/030	Item	289.86
CRABTREE POLESTAR SINGLE POLE MCBS; 20A SINGLE POLE TYPE-D 60D20	Item	10.21
CRABTREE POLESTAR SINGLE POLE MCBS; 40A SINGLE POLE TYPE-D 60D40	Item	10.77
CRABTREE POLESTAR SINGLE POLE MCBS; 50A SINGLE POLE TYPE-D 60D50	Item	10.77
CRABTREE POLESTAR SINGLE POLE MCBS; 63A SINGLE POLE TYPE-C 60C63	Item	10.48
CRABTREE POLESTAR TRIPLE POLE MCBS; 16A TRIPLE POLE TYPE-D 63D16	Item	42.97
CRABTREE POLESTAR TRIPLE POLE MCBS; 20A TRIPLE POLE TYPE-C 63C20	Item	41.76

© NSR 2008-2009

BASIC PRICES : LABOUR, PLANT AND MATERIALS

V:ELECTRICAL SERVICES : WORKS OF ALTERATION/SMALL WORKS
DISTRIBUTION EQUIPMENT

DISTRIBUTION BOARDS - (MCB)

Description	Unit	Price
CRABTREE POLESTAR TRIPLE POLE MCBS; 50A TRIPLE POLE TYPE-C 63C50	Item	44.14
CRABTREE POLESTAR TRIPLE POLE MCBS; 50A TRIPLE POLE TYPE-D 63D50	Item	45.42
CRABTREE POLESTAR TRIPLE POLE MCBS; 63A TRIPLE POLE TYPE-C 63C63	Item	44.14
CRABTREE POLESTAR TRIPLE POLE MCBS; 6A TRIPLE POLE TYPE-C 63C06	Item	44.14
CRABTREE STARBREAKER RCBOS; 20A 30MA SINGLE POLE RCBO TYPE C 602C/203	Item	68.46
CRABTREE STARBREAKER RCBOS; 32A 30MA SINGLE POLE RCBO TYPE C 602C/323	Item	68.46
CRABTREE STARBREAKER RCBOS; 40A 30MA SINGLE POLE RCBO TYPE C 602C/403	Item	68.46
DORMAN SMITH SWITCHGEAR LOADLIMITER MODULAR DEVICES; CONTACTOR 24A + 3NO CONTACTS LLM3P24C	Item	30.02
DORMAN SMITH SWITCHGEAR LOADLIMITER MODULAR DEVICES; CONTACTOR 24A + 4NO CONTACTS LLM4P24C	Item	32.21
DORMAN SMITH SWITCHGEAR LOADLIMITER MODULAR DEVICES; CONTACTOR 40A + 2NO CONTACTS LLM2P40C	Item	36.12
DORMAN SMITH SWITCHGEAR LOADLIMITER MODULAR DEVICES; CONTACTOR 63A + 4NO CONTACTS LLM4P63C	Item	89.90
DORMAN SMITH SWITCHGEAR LOADLIMITER MODULAR DEVICES; CONTACTOR 63A 3P + 3 N/O CONTACTS LLM3P63C - DISCONTINUED	Item	69.57
HAGER MCB/RCDS; 10A 30MA MCB/RCD AD105	Item	70.52
HAGER MCB/RCDS; 16A 30MA MCB/RCD AD107	Item	70.52
HAGER MCB/RCDS; 20A 30MA MCB/RCD AD108	Item	70.52
HAGER MCB/RCDS; 32A 30MA MCB/RCD AD110	Item	70.52
HAGER MCB/RCDS; 40A 30MA MCB/RCD AD111	Item	70.52
MERLIN GERIN RCDS; 100A 30MA RCD DP RMG1000302	Item	101.96
MERLIN GERIN RCDS; 20A 30MA COMBINED MCB/RCD 19666	Item	71.33
MERLIN GERIN RCDS; 40A 30MA COMBINED MCB/RCD 19669	Item	73.43
MERLIN GERIN RCDS; 63A 30MA VIGI RCD 4POLE MGV63 030 4	Item	91.42
MERLIN GERIN RCDS; 63A 30MA VIGI RCD DP MGV63 030 2	Item	62.76
SQUARE D TRIPLE POLE MCBS; 100A TRIPLE POLE TYPE C MCB KQ10C3100	Item	83.78
SQUARE D TRIPLE POLE MCBS; 80A TRIPLE POLE QOH MCB KQ10C380	Item	81.16
WYLEX RCDS - FOUR POLE; 63A 30MA 4POLE RCD WRS63/4	Item	90.78

BASIC PRICES : LABOUR, PLANT AND MATERIALS

V:ELECTRICAL SERVICES : WORKS OF ALTERATION/SMALL WORKS
DISTRIBUTION EQUIPMENT

FUSE HOLDERS AND BASES/FUSE WIRE

MEM FUSE CARRIERS; 100AMP CARRIER + HRC FUSE 10SCHF	Item	37.01
MEM FUSE CARRIERS; 10AMP CARRIER + FUSE 1SCHF	Item	6.98
MEM FUSE CARRIERS; 20AMP CARRIER + HRC FUSE 2SCHF	Item	6.30
MEM FUSE CARRIERS; 32AMP CARRIER + HRC FUSE 3SCHF	Item	9.19
MEM FUSE CARRIERS; 63AMP CARRIER + HRC FUSE 6SCHF	Item	16.49

FUSE LINKS (HRC)

ACEL FUSE LINKS; 100AMP BS88 OFF-SET TAGS, 2 HOLE FIXING AC3347	Item	23.75
ACEL FUSE LINKS; 20AMP BS88 OFF SET TAGS, 2 HOLE FIXING AC3304	Item	6.45
ACEL FUSE LINKS; 40AMP BS88 OFFSET TAGS, 2 HOLE FIXING AC3330	Item	12.95
ACEL FUSE LINKS; 63AMP BS88 OFFSET TAGS, 2 HOLE FIXING AC3332	Item	13.70
BUSSMANN CENTRE BOLTED FUSE LINKS; 50AMP BS88 2 HOLE FIXING AAO32M50	Item	4.62
MEM FUSE LINKS; 10A BS1362 250V 1510LC	Item	0.99
MEM FUSE LINKS; 5A BS646 250V 5LC	Item	0.75
MEM HOUSE SERVICE CUT-OUT FUSE LINKS; 40AMP FERRULE CAP BS1361 404R	Item	4.20
MEM HOUSE SERVICE CUT-OUT FUSE LINKS; 45AMP FERRULE CAP BS1361 454R	Item	4.20
MEM PARAMOUNT BOLTED CENTRE CONTACTS FUSE LINKS; 100AMP BS88 2 HOLE FIXING 100SF5	Item	10.31
MEM PARAMOUNT BOLTED OFF-SET CONTACTS FUSE LINKS; 63AMP BS88 2 HOLE FIXING 63SB4	Item	6.68
MEM PARAMOUNT BOLTED OFF-SET CONTACTS FUSE LINKS; 80AMP BS88 2 HOLE FIXING 80SD5	Item	10.21
MEM PARAMOUNT CONSUMER UNIT FUSE LINKS; 15AMP BLUE FERRULE CAP BS1361 15LC	Item	0.99
MEM PARAMOUNT CONSUMER UNIT FUSE LINKS; 30AMP RED FERRULE CAP BS1361 30LC	Item	1.11

MOULDED CASE (MCCB) CIRCUIT BREAKERS AND PANELBOARDS

CRABTREE POWERSTAR MCCBS; 100A SINGLE POLE 7PBJN1100	Item	65.00
CRABTREE POWERSTAR MCCBS; 100A TRIPLE POLE 7PBJN3100	Item	140.00
CRABTREE POWERSTAR MCCBS; 16A SINGLE POLE 7PBJN116	Item	60.00
CRABTREE POWERSTAR MCCBS; 32A TRIPLE POLE 7PBJN332	Item	130.00

BASIC PRICES : LABOUR, PLANT AND MATERIALS

V:ELECTRICAL SERVICES : WORKS OF ALTERATION/SMALL WORKS
DISTRIBUTION EQUIPMENT

MOULDED CASE (MCCB) CIRCUIT BREAKERS AND PANELBOARDS

Description	Unit	Price
CRABTREE POWERSTAR MCCBS; 50A SINGLE POLE 7PBJN150	Item	65.00
CRABTREE POWERSTAR MCCBS; 50A TRIPLE POLE 7PBJN350	Item	130.00
CRABTREE POWERSTAR MCCBS; 80A SINGLE POLE 7PBJN180	Item	65.00

RESIDUAL CURRENT DEVICES

Description	Unit	Price
DORMAN SMITH RCDS; 40A 30MA 4POLE RCCB LLMA40/30/4	Item	105.30
MEM MEMSHIELD RCDS; 100A 30MA 4POLE RCD AM1004H	Item	135.64
MEM MEMSHIELD RCDS; 25A 30MA 2POLE RCD A25HE	Item	58.70
MEM MEMSHIELD RCDS; 4 MODULE MOULDED ENCLOSURE AN4EBLS	Item	11.01
MEM MEMSHIELD RCDS; 40A 30MA 2POLE RCD AM40HE	Item	63.88
MEM MEMSHIELD RCDS; 40A 30MA 4POLE RCD AM404H	Item	82.46
MEM MEMSHIELD RCDS; 80A 30MA 2POLE RCD A80HE	Item	83.39
MEM MEMSHIELD RCDS; RCD ENCLOSURE 2 MODULE IP30 AM1SL	Item	7.22
MERLIN GERIN RCDS; 25A 30MA 4POLE RCD RMG250304	Item	74.07

SWITCH FUSES 500V

Description	Unit	Price
MEM - EXEL RANGE SWITCH FUSE; 100A DP + HRC FUSES 100KXDC2F	Item	309.39
MEM - EXEL RANGE SWITCH FUSE; 100A TPN + HRC FUSE 100KXTNC2F	Item	407.09
MEM - EXEL RANGE SWITCH FUSE; 20A DP + HRC FUSE 15KXDC2F	Item	87.69
MEM - EXEL RANGE SWITCH FUSE; 20A TPN + HRC FUSE 15KXTNC2F	Item	106.15
MEM - EXEL RANGE SWITCH FUSE; 63A DP + HRC FUSE 60KXDC2F	Item	180.96
MEM - EXEL RANGE SWITCH FUSE; 63A TPN + HRC FUSE 60KXTNC2F	Item	232.40

SWITCH ISOLATORS 500V

Description	Unit	Price
MEM - EXEL RANGE SWITCH ISOLATOR; 100A DP SWITCH 100AXD2	Item	178.50
MEM - EXEL RANGE SWITCH ISOLATOR; 100A TPN SWITCH 100AXTN2	Item	235.75
MEM - EXEL RANGE SWITCH ISOLATOR; 125A TPN SWITCH 125AXTN2	Item	254.42
MEM - EXEL RANGE SWITCH ISOLATOR; 20A DP SWITCH 15AXD2	Item	67.82

BASIC PRICES : LABOUR, PLANT AND MATERIALS

V:ELECTRICAL SERVICES : WORKS OF ALTERATION/SMALL WORKS
DISTRIBUTION EQUIPMENT
SWITCH ISOLATORS 500V

MEM - EXEL RANGE SWITCH ISOLATOR; 20A TPN SWITCH 15AXTN2	Item	77.88
MEM - EXEL RANGE SWITCH ISOLATOR; 32A DP SWITCH 30AXD2	Item	83.00
MEM - EXEL RANGE SWITCH ISOLATOR; 32A TPN SWITCH 30AXTN2	Item	92.30
MEM - EXEL RANGE SWITCH ISOLATOR; 63A DP SWITCH 60AXD2	Item	117.39
MEM - EXEL RANGE SWITCH ISOLATOR; 63A TPN SWITCH 60AXTN2	Item	149.46

ELECTRIC MOTORS
ELECTRIC MOTORS - THREE PHASE

3 PHASE 50HZ ALUMINIUM. FOOT & FLANGE MOUNTING. FRAMES 63-250. 0.18KW-55KW; D100 3KW FLANGE MTG 3PHASE 50HZ ALUM MOTOR D100-3KW1500	Item	290.46
3 PHASE 50HZ ALUMINIUM. FOOT & FLANGE MOUNTING. FRAMES 63-250. 0.18KW-55KW; D112 4KW FOOT MTG 3PHASE 50HZ ALUM MOTOR D112-4KW3000FM	Item	340.81
3 PHASE 50HZ ALUMINIUM. FOOT & FLANGE MOUNTING. FRAMES 63-250. 0.18KW-55KW; D132 5.5KW FLANGE MTG 3PHASE 50HZ ALUM MOTOR D132-5.5KW3000	Item	466.58
3 PHASE 50HZ ALUMINIUM. FOOT & FLANGE MOUNTING. FRAMES 63-250. 0.18KW-55KW; D160 7.5KW FOOT MTG 3PHASE 50HZ ALUM MOTOR D160-7.5KW1000FM	Item	808.48
3 PHASE 50HZ ALUMINIUM. FOOT & FLANGE MOUNTING. FRAMES 63-250. 0.18KW-55KW; D63 0.25KW FLANGE MTG 3PHASE 50HZ ALUM MOTOR D63-.25KW3000	Item	116.26
3 PHASE 50HZ ALUMINIUM. FOOT & FLANGE MOUNTING. FRAMES 63-250. 0.18KW-55KW; D80 1.1KW FLANGE MTG 3PHASE 50HZ ALUM MOTOR D80-1.1KW3000	Item	188.22
3 PHASE 50HZ ALUMINIUM. FOOT & FLANGE MOUNTING. FRAMES 63-250. 0.18KW-55KW; D90 2.2KW FLANGE MTG 3PHASE 50HZ ALUM MOTOR D90-2.2KW3000	Item	249.41

GENERAL
Sundries

Materials; Unit material rate	BASE	1.35

GRID SWITCH SYSTEMS
CRABTREE

SWITCHES AND ACCESSORIES; 20A 1 WAY SWITCH - WHITE 4450WH	Item	2.93
SWITCHES AND ACCESSORIES; 20A 2 WAY + OFF SWITCH - WHITE 4552WH	Item	4.85

BASIC PRICES : LABOUR, PLANT AND MATERIALS

V:ELECTRICAL SERVICES : WORKS OF ALTERATION/SMALL WORKS

GRID SWITCH SYSTEMS

CRABTREE

SWITCHES AND ACCESSORIES; 20A 2 WAY KEY OPERATED SWITCH - WHITE 4551WH	Item	6.68
SWITCHES AND ACCESSORIES; 20A 2 WAY SWITCH - WHITE 4550WH	Item	3.81
SWITCHES AND ACCESSORIES; 20A DP 1WAY SWITCH - WHITE 4011/1	Item	7.04
SWITCHES AND ACCESSORIES; 20A DP KEY OPERATED SWITCH - WHITE 4461WH	Item	6.73
SWITCHES AND ACCESSORIES; 20A DP SWITCH - WHITE 4460WH	Item	4.29
SWITCHES AND ACCESSORIES; 20A INTERMEDIATE SWITCH - WHITE 4535WH	Item	7.12
SWITCHES AND ACCESSORIES; 3 GANG PLATE - SATIN CHROME 7B03/SC	Item	7.61
SWITCHES AND ACCESSORIES; 4 GANG PLATE - SATIN CHROME 7B04/SC	Item	7.61
SWITCHES AND ACCESSORIES; 6 GANG PLATE - SATIN CHROME 7B06/SC	Item	13.95
SWITCHES AND ACCESSORIES; 6A RETRACTIVE SWITCH - WHITE 4489	Item	4.92
SWITCHES AND ACCESSORIES; BELL PUSH WHITE 4490WH	Item	5.08
SWITCHES AND ACCESSORIES; BLANK COMPONENT 4492WH	Item	1.52
SWITCHES AND ACCESSORIES; FUSE UNIT 13AMP 4436WH	Item	8.27

MK

EDGE GRID PLUS SWITCH MODULES - 20AMP - NEW RANGE; 20A DP 1WAY SECRET KEY SWITCH K4917 BLK	Item	6.80
EDGE GRID PLUS SWITCH MODULES - 20AMP - NEW RANGE; 20A INTERMEDIATE SWITCH BRUSHED STAINLESS STEEL C/W WHITE INSERTS K4893 BSS	Item	12.96
EDGE GRID PLUS SWITCH MODULES - 20AMP - NEW RANGE; 20A SP 1WAY SWITCH BRUSHED STAINLESS STEEL C/W WHITE INSERTS K4891 BSS	Item	7.08
EDGE GRID PLUS SWITCH MODULES - 20AMP - NEW RANGE; 20A SP 2WAY SECRET KEY SWITCH K4898 BLK	Item	7.26
EDGE GRID PLUS SWITCH MODULES - 20AMP - NEW RANGE; 20A SP 2WAY SWITCH BRUSHED STAINLESS STEEL C/W WHITE INSERTS K4892 BSS	Item	7.89

HAZARDOUS AREA PRODUCTS

BICC COMPONENTS

GLAND EARTH TAGS - BRASS; BRASS EARTH TAG 32MM AL0048	Item	0.75

BASIC PRICES : LABOUR, PLANT AND MATERIALS

V:ELECTRICAL SERVICES : WORKS OF ALTERATION/SMALL WORKS
HEATING CONTROL
TIMESWITCHES

<NONE>; 16022 COMPACT QUARTZ TIME SWITHC SINGLE CHANNEL 24HR	nr	30.57
<NONE>; 16721 COMPACT QUARTZ TIME SWITCH SINGLE CHANNEL 7DAY	nr	38.16
<NONE>; E854 ELECTRONIC TIMER ROUND PATTERN 3 PIN 7DAY	nr	171.57
<NONE>; E855 ELECTRONIC TIMER ROUND PATTERN 4 PIN 7DAY	nr	174.45
<NONE>; EP2002 PROGRAMMER 5/2DAY	EACH	78.29
<NONE>; Q554 FORM 2 TIME SWITCH ROUND PATTERN 3 PIN 24HR	nr	125.38
<NONE>; Q555 FORM 2 TIME SWITCH ROUND PATTERN 4 PIN 24HR	nr	130.29

INDUSTRIAL CONTROL & AUTOMATION PRODUCTS
CONTACTORS, STARTERS, RELAYS AND OVERLOADS

ACEL 3POLE AC OPERATION BLOCK CONTACTORS; 12A AC3 5.5KW CONTACTOR 230/240V 50HZ 1NC CONTACT AC3400	Item	16.20
ACEL 3POLE AC OPERATION BLOCK CONTACTORS; 17A AC3 7.5KW CONTACTOR 24V 1NO CONTACT AC3401	Item	20.50
ACEL 3POLE AC OPERATION BLOCK CONTACTORS; 37A AC3 18.5KW CONTACTOR 230/240V 50HZ 1NO CONTACT AC3404	Item	52.95
ACEL 3POLE AC OPERATION BLOCK CONTACTORS; C23 240V 22A 3POLE AC3 50HZ AC3402	Item	29.15
ACEL 3POLE AC OPERATION BLOCK CONTACTORS; C3 240V 5.7A 3POLE AC3 50HZ AC3400	Item	16.20
ACEL 3POLE AC OPERATION BLOCK CONTACTORS; C33 415V 28A 3POLE AC3 50HZ AC3403	Item	43.20
CRABTREE MOTORPAK DIRECT-ON-LINE STARTERS; D-O-L STARTER 240V 28ADS1X	Item	41.27
CRABTREE MOTORPAK DIRECT-ON-LINE STARTERS; D-O-L STARTER 415V 48ADS1X	Item	43.40
CRABTREE MOTORPAK OVERLOAD RELAYS; OVERLOAD RELAY 0.60-1.0A RANGE 55500/FC	Item	24.28
CRABTREE MOTORPAK OVERLOAD RELAYS; OVERLOAD RELAY 15.0-25A RANGE 57500/PC	Item	31.24
CRABTREE MOTORPAK OVERLOAD RELAYS; OVERLOAD RELAY 2.4-4.0A RANGE 555000/JC	Item	24.28
CRABTREE MOTORPAK OVERLOAD RELAYS; OVERLOAD RELAY 3.8-6.3A RANGE 55500/KC	Item	24.28
CRABTREE MOTORPAK OVERLOAD RELAYS; OVERLOAD RELAY 6.0-10.0A RANGE 555000/MC	Item	25.58
CRABTREE MOTORPAK OVERLOAD RELAYS; OVERLOAD RELAY 9.6-16.0A RANGE 55500/NC	Item	28.31

BASIC PRICES : LABOUR, PLANT AND MATERIALS

V:ELECTRICAL SERVICES : WORKS OF ALTERATION/SMALL WORKS
INDUSTRIAL CONTROL & AUTOMATION PRODUCTS

CONTACTORS, STARTERS, RELAYS AND OVERLOADS

Description	Unit	Price
CRABTREE OPEN TYPE CONTACTORS; CEICON CONTACTOR 240V 50HZ 48400/XB	Item	88.20
CRABTREE OPEN TYPE CONTACTORS; CEICON CONTACTOR 240V 50HZ 49400/XB	Item	122.12
CRABTREE STAR-DELTA OVERLOADS; STARTER STAR DELTA IP65 11KW MSDS11 - NO LONGER USED	Item	357.97
CRABTREE STAR-DELTA OVERLOADS; STARTER STAR DELTA IP65 5.5KW MSDS5.5 - NO LONGER USED	Item	336.23
MEM AUTOLINE HEATING AND LIGHTING DUTY CONTACTORS; 20A 7.5KW 3POLE 220/240V COIL 2812004VCOA	Item	31.78
MEM AUTOLINE HEATING AND LIGHTING DUTY CONTACTORS; 32A 11KW 3POLE 220/240V COIL 4825004VCOA	Item	41.07
NEWLEC OVERLOADS; 1.0 - 1.4 OVERLOAD - NLOLB1	nr	23.90
NEWLEC OVERLOADS; 1.3A - 5.0 OVERLOAD - NLOLB2 - 6	nr	24.20
NEWLEC OVERLOADS; 4.4A - 8.5A OVERLOAD - NLOLB7 - 8	nr	27.00
NEWLEC OVERLOADS; 7.5A - 19A OVERLOAD - NLOLB9 - 11	nr	28.60
NEWLEC STAR-DELTA STARTERS; CONTACTOR 400V 50HZ 24A NLSD415B	Item	211.55
PROTEUS CONTACTORS; C93 415V 80A 3POLE AC3 50HZ HLC80	Item	187.89

CONTROL AND SIGNALLING (PUSHBUTTONS)

Description	Unit	Price
CRABTREE ALL PURPOSE PUSHBUTTON UNITS; PUSHBUTTON UNIT FORWARD/REVERSE/STOP 3MB	Item	36.92
CRABTREE ALL PURPOSE PUSHBUTTON UNITS; STOP PUSHBUTTON UNIT MUSHROOM HEAD 1/MBM	Item	18.56
CRABTREE ALL PURPOSE PUSHBUTTON UNITS; STOP PUSHBUTTON UNIT. AUTOLOCK. MUSHROOM HEAD 1/MBA	Item	24.10
CRABTREE ALL PURPOSE PUSHBUTTON UNITS; STOP/START PUSHBUTTON UNIT. AUTOLOCK. MUSHROOM HEAD 2MBA	Item	33.15
CRABTREE ASSEMBLED PUSH BUTTON UNITS; PUSHBUTTON STATION AUTOLOCK STOP 22931/AV	Item	37.57
CRABTREE ASSEMBLED PUSH BUTTON UNITS; PUSHBUTTON STATION AUTOLOCK STOP 1/2 COLLAR SHROUD 22931/AHCV	Item	50.94

ENCLOSURES FOR PUSH BUTTON AND INDICATORS

Description	Unit	Price
CRABTREE ENCLOSURES; PLASTIC ENCLOSURE 1 GANG GREY COVER IP65 22EV1	Item	7.40
CRABTREE ENCLOSURES; PLASTIC ENCLOSURE 1 GANG YELLOW COVER IP65 22EVG1	Item	7.40

BASIC PRICES : LABOUR, PLANT AND MATERIALS

V:ELECTRICAL SERVICES : WORKS OF ALTERATION/SMALL WORKS
INDUSTRIAL CONTROL & AUTOMATION PRODUCTS
INDICATOR LIGHT UNITS

INDICATING LIGHTS; INDICATOR LIGHT C/W E10 NEON LAMP SVN125	Item	6.88
INDICATING LIGHTS; LENS RED 22.5MM HAGER SW082	Item	0.82

RAIL MOUNTED TERMINALS

NEWLEC RAIL MOUNTED TERMINALS; 2.5MM TERMINAL 10 PACK (FOR 32MM G-RAIL MOUNTING) NLTERM2.5/32	Item	14.00
NEWLEC RAIL MOUNTED TERMINALS; 32MM G-RAIL 500MM LENGTH NLRAIL32	Item	2.10
NEWLEC RAIL MOUNTED TERMINALS; 4.0MM TERMINAL 10 PACK (FOR 32MM G-RAIL MOUNTING) NLTERM4.0/32	Item	11.95

ROTARY SWITCHES

CRABTREE MOTOR CIRCUIT SWITCH; 25A TP + NL MOTOR CIRCUIT SWITCH INSULATED IP54 15425/12	Item	16.31

LAMPS & TUBES
COMPACT FLUORESCENT LAMPS

GE LIGHTING - 2D COMPACT LAMPS - POLYLUX; 2PIN 2700K WARM WHITE 28W	Item	6.75
PHILIPS PLCE (PLEC) ELECTRONIC RANGE COMPACT LAMPS; PLF18830 18W 4 PIN	Item	7.88
PHILIPS SL COMPACT LAMPS; 16W ES PRISMATIC SL16ES	Item	5.87

DISPLAY LAMPS

NEWLEC REFLECTOR SPOT LAMPS - NEW RANGE; R80 100W ES 240V NLR80-100	Item	0.69

FLUORESCENT TUBES

FLUORESCENT TUBES - MINIATURE - T5 (16MM); WHITE L823 12"8W	Item	1.35
FLUORESCENT TUBES - T12 (38MM); COOL WHITE F100W/33	Item	6.25
FLUORESCENT TUBES - T8 (26MM); WARM WHITE F18W/29 2'18W	Item	2.05
FLUORESCENT TUBES - T8 (26MM); WARM WHITE F30W/29 3'30W	Item	3.05
FLUORESCENT TUBES - T8 (26MM); WARM WHITE F36W/29 4'36W	Item	2.10
FLUORESCENT TUBES - T8 (26MM); WARM WHITE F58W/29 5'58W	Item	2.35
FLUORESCENT TUBES - T8 (26MM); WARM WHITE F70W/29 6'70W	Item	3.95
U SHAPED FLUORESCENT TUBES; WHITE F70W/29 40W	Item	7.28

MERCURY DISCHARGE LAMPS

BASIC PRICES : LABOUR, PLANT AND MATERIALS

V:ELECTRICAL SERVICES : WORKS OF ALTERATION/SMALL WORKS
LAMPS & TUBES

MERCURY DISCHARGE LAMPS

MERCURY DISCHARGE (MBFU/MBFR/MBFT) LAMPS; 3PIN BC MBF 125W	Item	31.63
MERCURY DISCHARGE (MBFU/MBFR/MBFT) LAMPS; ES MLL (MBTF) BLENDED 160W HMLI160	Item	93.42
MERCURY DISCHARGE (MBFU/MBFR/MBFT) LAMPS; GES HPLN (MBFU) 1000W	Item	275.64
MERCURY DISCHARGE (MBFU/MBFR/MBFT) LAMPS; GES HPLN (MBFU) 700W	Item	211.43
MERCURY DISCHARGE (MBFU/MBFR/MBFT) LAMPS; GES MBF H250/40 250W	Item	77.65
MERCURY DISCHARGE (MBFU/MBFR/MBFT) LAMPS; GES MBF H400/40 400W	Item	124.23
MERCURY DISCHARGE (MBFU/MBFR/MBFT) LAMPS; GES MBTF HSBBW O020478 500W	Item	148.67

SODIUM DISCHARGE (SON) LAMPS

GE LIGHTING HIGH PRESSURE - LUCALOX RANGE LAMPS; LAMP SODIUM SON-E E40 GES LU150/100/D/40 150W	Item	119.26
GE LIGHTING HIGH PRESSURE - LUCALOX RANGE LAMPS; LAMP SODIUM SON-E E40 GES LU250/D/40 250W	Item	163.05
GE LIGHTING HIGH PRESSURE - LUCALOX RANGE LAMPS; LAMP SODIUM SON-E GES LU400/D/40 400W	Item	209.66
GE LIGHTING HIGH PRESSURE - LUCALOX RANGE LAMPS; LAMP SODUIM SON-T E27 ES LU50/90/T/27 50W	Item	97.91
GE LIGHTING HIGH PRESSURE - LUCALOX RANGE LAMPS; RX7S (DOUBLE ENDED) 1000W SONTD	Item	473.88
OSRAM VIALOX HIGH PRESSURE - (SON-E/SON-T/SON-PT) LAMPS; NAVE (SONE) EXTERNAL S70 70W	Item	19.85
OSRAM VIALOX HIGH PRESSURE - (SON-E/SON-T/SON-PT) LAMPS; NAVE (SONE) INTERNAL 100W	Item	43.77
OSRAM VIALOX HIGH PRESSURE - (SON-E/SON-T/SON-PT) LAMPS; NAVE (SONE) INTERNAL SONT400I 400W	Item	45.00
OSRAM VIALOX HIGH PRESSURE - (SON-E/SON-T/SON-PT) LAMPS; NAVT (SONT) EXTERNAL 400W	Item	42.50
OSRAM VIALOX HIGH PRESSURE - (SON-E/SON-T/SON-PT) LAMPS; NAVT (SONT) INTERNAL 1000W	Item	65.00
OSRAM VIALOX HIGH PRESSURE - (SON-E/SON-T/SON-PT) LAMPS; NAVT (SONT) INTERNAL 250W	Item	35.50

SODIUM DISCHARGE (SOX) LAMPS

LOW PRESSURE (SOX/SOX-E) LAMPS; 36W BC SOX-E	Item	16.82
LOW PRESSURE (SOX/SOX-E) LAMPS; BC SOX SX135 135W	Item	36.60
LOW PRESSURE (SOX/SOX-E) LAMPS; BC SOX SX35 35W	Item	19.95
LOW PRESSURE (SOX/SOX-E) LAMPS; BC SOX SX55 55W	Item	23.35

BASIC PRICES : LABOUR, PLANT AND MATERIALS

V:ELECTRICAL SERVICES : WORKS OF ALTERATION/SMALL WORKS

LAMPS & TUBES

SODIUM DISCHARGE (SOX) LAMPS

LOW PRESSURE (SOX/SOX-E) LAMPS; BC SOX-E ECONOMY 135W	Item	22.08
LOW PRESSURE (SOX/SOX-E) LAMPS; BC SOX-E ECONOMY SXE66 66W	Item	20.55

TUNGSTEN FILAMENT LAMPS

GLS LAMPS - 240V; 240V BC CLEAR 100W	Item	0.57
NEWLEC GLS LAMPS - NEW RANGE; 100W BC PEARL 240V (PACK OF 2) NLGLS100	Item	0.59

TUNGSTEN HALOGEN LAMPS (LOW VOLTAGE)

DICHROIC REFLECTOR LAMPS; 35W 12V 36° GU5.3 CLOSED FACE M681	Item	3.95

TUNGSTEN HALOGEN LAMPS (MAINS VOLTAGE)

HALOGENA LAMPS; 240V 100W R7S LINEAR HALOGEN	Item	1.35
HALOGENA LAMPS; 240V 150W R7S LINEAR HALOGEN	Item	1.35

LIGHTING FITTINGS & EMERGENCY LIGHTING

AMENITY/SECURITY LIGHTING

COUGHTRIE QUORUM RANGE; 16W 2D BULKHEAD. BLACK BASE, POLYCARBONATE DIFFUSER CPS16/B/CL	Item	45.30

COMPACT FLUORESCENT LUMINAIRES

NEWLEC DRUM LUMINAIRE; CIRCULAR FITTING + 2D 16W LAMP + POLYPROPYLENE DIFFUSER NLDRM16WC	Item	19.80
NEWLEC DRUM LUMINAIRE; SQUARE FITTING + 2D 28W LAMP + POLYCARBONATE DIFFUSER NLDRM28WSQ	Item	28.10
NEWLEC LOW ENERGY BULKHEAD; WHITE BASE CIRCULAR 2D 38W IP65 NL3771	Item	35.20

CONTROL GEAR AND SPARES

ACEL LIGHTING CAPACITORS; 25MFD CAPACITORS 250V AC4987	Item	5.10
ACEL LIGHTING CAPACITORS; 4MFD CAPACITOR AC4980	Item	4.30
FLUORESCENT HIGH FREQUENCY COMPACT BALLASTS (NON DIMMABLE); 55W BALLAST COMPACT ELECTRONIC HFP255LLE11	Item	60.40
FLUORESCENT HIGH FREQUENCY T8/T12 BALLASTS (NON DIMMABLE); 2X36W TYPE-E BALLAST HF ELECTRONIC HFP236PLLE11	Item	60.40
FLUORESCENT SWITCH-START CHOKES; 100W CHOKE T12 LINEAR AC4918	Item	9.95
FLUORESCENT SWITCH-START CHOKES; 18W CHOKE T8 AC4913	Item	6.65

BASIC PRICES : LABOUR, PLANT AND MATERIALS

V:ELECTRICAL SERVICES : WORKS OF ALTERATION/SMALL WORKS
LIGHTING FITTINGS & EMERGENCY LIGHTING

CONTROL GEAR AND SPARES

FLUORESCENT SWITCH-START CHOKES; 30W CHOKE T8 AC4914	Item	7.20
FLUORESCENT SWITCH-START CHOKES; 36W CHOKE T8 AC4915	Item	7.20
GE LIGHTING FLUORESCENT STARTER SWITCH; STARTER SWITCH 70/100W 155/801	Item	4.63
LAMPHOLDERS (MISCELLANEOUS); BATTEN LAMPHOLDER E10 SCREW FIX GB1665	Item	2.40
LAMPHOLDERS (MISCELLANEOUS); BI-PIN LAMPHOLDER 2.54MM PCB SOLDER GB1598	Item	3.13
LAMPHOLDERS (MISCELLANEOUS); LAMPHOLDER CERAMIC E27 (ES) GL1131	Item	6.26
MERCURY LAMP BALLASTS; 400W MBF MERCURY BALLAST NLSON400V	Item	44.85
NEWLEC HIGH PRESSURE SODIUM (SON) BALLASTS - NEW RANGE; BALLAST FOR SON LAMPS 250W NLSON250	Item	32.75
SOX (LOW PRESSURE SODIUM DISCHARGE) LAMP IGNITORS; 66W SOX IGNITOR SX70	Item	46.81
THORN LIGHTING CONTROL GEAR AND SPARES; CAPACITOR 10UF GC2442	Item	8.40
TRANSTAR FLUORESCENT CONTROL GEAR; 65W/85W HPF ELECTRONIC BALLAST K85XE	Item	36.73

DIMMER CONTROLS

DIMMERS; 250W 2 GANG 2WAY. PUSH WHITE GD2G2W	Item	22.83
DIMMERS; 400W 1 GANG 1WAY. PUSH WHITE GD1G1W4	Item	14.20

EMERGENCY LIGHTING

CHLORIDE EMERGENCY LUMINAIRES; 2X18W TWINLIGHT NON-MAINTAINED 3HOUR SURTWIN	Item	73.68
CHLORIDE EMERGENCY LUMINAIRES; CLASSIC 8 LUMINAIRE 8W FLUORESCENT NON-MAINTAINED 3HOUR CA8/NM/3FS2	Item	20.83
CHLORIDE EMERGENCY LUMINAIRES; SQUARE 28W 2D FLUORESCENT MAINTAINED 3HOUR - IP65 DRG283M23024	Item	142.90
CHLORIDE EMERGENCY LUMINAIRES; TUNGSTEN HALOGEN TWINLIGHT 10W 3HOUR WPTP10NM3/TH	Item	249.17
CHLORIDE EMERGENCY LUMINAIRES; ZENITH 8W FLUORESCENT MAINTAINED ZEN8M3F/P	Item	103.73

FLOOD LIGHTING

NEWLEC FLOODLIGHT - TUNGSTEN HALOGEN; 1000W ENCLOSED FLOODLIGHT NL3915	Item	25.00
NEWLEC FLOODLIGHT - TUNGSTEN HALOGEN; 500W PROFESSIONAL QUALITY ENCLOSED FLOODLIGHT NL500PRO	Item	9.55

BASIC PRICES : LABOUR, PLANT AND MATERIALS

V:ELECTRICAL SERVICES : WORKS OF ALTERATION/SMALL WORKS
LIGHTING FITTINGS & EMERGENCY LIGHTING

FLOOD LIGHTING

NEWLEC FLOODLIGHTS AND WALLWASHERS - DISCHARGE; 150W SON FLOODLIGHT C/W 150W SON LAMP - IP65 NL3795	Item	94.05
NEWLEC FLOODLIGHTS AND WALLWASHERS - DISCHARGE; 250W SON FLOODLIGHT C/W 250W SON LAMP - IP65 NL3796	Item	96.84
NEWLEC FLOODLIGHTS AND WALLWASHERS - DISCHARGE; 70W SON FLOODLIGHT C/W 70W SON LAMP - IP65 NL3793	Item	37.75
NEWLEC FLOODLIGHTS AND WALLWASHERS - DISCHARGE; 70W SON WALLWASHER + PHOTOCELL C/W 70W SON LAMP - IP55 NL3798	Item	129.00
THORN LIGHTING SONPAK FLOODLIGHT; SONPAK 15 + LAMP 150W SON OTX150	Item	153.12
THORN LIGHTING SONPAK FLOODLIGHT; SONPAK 25 + LAMP 250W SON OTX250	Item	178.25
THORN LIGHTING SONPAK FLOODLIGHT; SONPAK 40 400W SODIUM E40 LAMP OTX400	Item	195.27
THORN LIGHTING SONPAK FLOODLIGHT; SONPAK 7 + 70W SON-E LAMP OTX70	Item	76.78

FLUORESCENT BATTENS AND ATTACHMENTS

CROMPTON LIGHTING - CROMPACK 5 OPAL DIFFUSER; 4FT SINGLE CPD41	Item	18.09
CROMPTON LIGHTING - CROMPACK 5 OPAL DIFFUSER; 4FT TWIN CPD42	Item	22.35
CROMPTON LIGHTING - CROMPACK 5 OPAL DIFFUSER; 5FT SINGLE CPD51	Item	20.09
CROMPTON LIGHTING - CROMPACK 5 OPAL DIFFUSER; 5FT TWIN CPD52	Item	24.51
CROMPTON LIGHTING - CROMPACK 5 OPAL DIFFUSER; 6FT SINGLE CPD61	Item	25.03
CROMPTON LIGHTING - CROMPACK 5 OPAL DIFFUSER; 6FT TWIN CPD62	Item	29.77
CROMPTON LIGHTING - CROMPACK 5 OPAL DIFFUSER; 8FT SINGLE CPD81	Item	32.40
CROMPTON LIGHTING - CROMPACK 5 OPAL DIFFUSER; 8FT TWIN CPD82	Item	41.62
THORN LIGHTING MK15 POPULAR PACK BATTEN + TUBE; 4FT SINGLE 36W + TUBE PP136	Item	27.17
THORN LIGHTING MK15 POPULAR PACK BATTEN + TUBE; 4FT TWIN 2X36W + TUBES PP236	Item	48.31
THORN LIGHTING MK15 POPULAR PACK BATTEN + TUBE; 5FT SINGLE 58W + TUBE PP158	Item	28.57
THORN LIGHTING MK15 POPULAR PACK BATTEN + TUBE; 5FT TWIN 2X58W + TUBES PP258	Item	51.62
THORN LIGHTING MK15 POPULAR PACK BATTEN + TUBE; 6FT SINGLE 70W + TUBE PP170	Item	33.72

BASIC PRICES : LABOUR, PLANT AND MATERIALS

V:ELECTRICAL SERVICES : WORKS OF ALTERATION/SMALL WORKS
LIGHTING FITTINGS & EMERGENCY LIGHTING

FLUORESCENT BATTENS AND ATTACHMENTS

Description	Unit	Price
THORN LIGHTING MK15 POPULAR PACK BATTEN + TUBE; 6FT TWIN 2X70W + TUBES PP270	Item	55.45
THORN LIGHTING MK15 POPULAR PACK BATTEN + TUBE; 8FT SINGLE 100W + TUBE PP1100	Item	77.47
THORN LIGHTING MK15 POPULAR PACK BATTEN + TUBE; 8FT TWIN 2X100W + TUBES PP2100	Item	138.27
THORN LIGHTING POPULAR PACK OPAL DIFFUSER; 4FT SINGLE PPD4	Item	18.40
THORN LIGHTING POPULAR PACK OPAL DIFFUSER; 4FT TWIN PPD24	Item	29.80
THORN LIGHTING POPULAR PACK OPAL DIFFUSER; 5FT SINGLE PPD5	Item	18.69
THORN LIGHTING POPULAR PACK OPAL DIFFUSER; 5FT TWIN PPD25	Item	31.41
THORN LIGHTING POPULAR PACK OPAL DIFFUSER; 6FT SINGLE PPD6	Item	24.00
THORN LIGHTING POPULAR PACK OPAL DIFFUSER; 6FT TWIN PPD26	Item	38.04
THORN LIGHTING POPULAR PACK OPAL DIFFUSER; 8FT SINGLE PPD8	Item	35.30
THORN LIGHTING POPULAR PACK OPAL DIFFUSER; 8FT TWIN PPD28	Item	57.00

LIGHTING CONTROL SYSTEMS

Description	Unit	Price
THORN LIGHTING CONTROL SYSTEM; PHOTOCELL SWITCH 6A SURFACE MOUNTED 801PCSF	Item	113.60

MAINS VOLTAGE DISPLAY LIGHTING

Description	Unit	Price
NEWLEC DOWNLIGHTERS - MAINS VOLTAGE (LESS LAMP); EYEBALL DOWNLIGHTER MAINS VOLTAGE CHROME FOR 100W R80 LAMP NLDL240R8CH	Item	3.40
REGGIANI LTD LIGHTSTREAM MAINS VOLTAGE DOWNLIGHTERS; 135MM RECESSED DOWNLIGHTER + 75W PAR30 SPOT LAMP - WHITE	Item	26.50
REGGIANI LTD LIGHTSTREAM MAINS VOLTAGE DOWNLIGHTERS; RECESSED EYEBALL DOWNLIGHTER + 50W PAR20 LAMP - WHITE	Item	30.00

RECESSED MODULAR FLUORESCENT LUMINAIRES

Description	Unit	Price
NEWLEC RECESSED MODULAR LUMINAIRES (ENCLOSED) - NEW RANGE; 1200X600MM 3X36W RECESSED MODULAR - SWITCH START NL5558N - DISCONTINUED	Item	25.73
NEWLEC RECESSED MODULAR LUMINAIRES (ENCLOSED) - NEW RANGE; 1200X600MM 4X36W RECESSED MODULAR - SWITCH START NL5516NS	Item	26.34
NEWLEC RECESSED MODULAR LUMINAIRES (ENCLOSED) - NEW RANGE; 1200X600MM PRISMATIC DIFFUSER	nr	3.55

BASIC PRICES : LABOUR, PLANT AND MATERIALS

V:ELECTRICAL SERVICES : WORKS OF ALTERATION/SMALL WORKS
LIGHTING FITTINGS & EMERGENCY LIGHTING

SURFACE FLUORESCENT LUMINAIRES

THORN LIGHTING ESCORT BULKHEAD + TUBE; 2X8W FLUORESCENT OBV2008	Item	52.67

TUNGSTEN LUMINAIRES

THORN LIGHTING BULKHEAD/WELLGLASS LUMINAIRE; TUNGSTEN BULKHEAD 100W OLG1100BC	Item	68.00
THORN LIGHTING BULKHEAD/WELLGLASS LUMINAIRE; TUNGSTEN BULKHEAD 2x26W THSOZ226C	Item	97.31

VAPOURPROOF FITTINGS IP65

ALTO FITTINGS; AL0923 BULKHEAD IP65 ROUND 16W/WHI	nr	20.75
ALTO FITTINGS; AL0927 BULKHEAD IP65 ROUND 28W/WHI	nr	31.10
NEWLEC FITTINGS; ANTI-CORROSIVE, POLYCARBONATE HIGH FREQUENCY 1270MM, 1X36W	nr	42.83
NEWLEC FITTINGS; ANTI-CORROSIVE, POLYCARBONATE HIGH FREQUENCY 1270MM, 2X36W	nr	48.63
NEWLEC FITTINGS; ANTI-CORROSIVE, POLYCARBONATE HIGH FREQUENCY 1570MM, 1X58W	nr	44.65
NEWLEC FITTINGS; ANTI-CORROSIVE, POLYCARBONATE HIGH FREQUENCY 1570MM, 2X58W	nr	52.80
NEWLEC FITTINGS; ANTI-CORROSIVE, POLYCARBONATE HIGH FREQUENCY 1850MM, 1X70W	nr	53.46
NEWLEC FITTINGS; ANTI-CORROSIVE, POLYCARBONATE HIGH FREQUENCY 1850MM, 2X70W	nr	62.73

SPACE HEATING

AIR CURTAINS

DIMPLEX AIR CURTAIN; 3KW WITH INTEGRAL CONTROLS AC3RN	Item	144.02
VENT-AXIA SCREEN ZONE; 3KW 3 WAY SWITCH - WARMAIR 3	Item	143.33

CONVECTOR HEATERS

DIMPLEX DX CONVECTORS - NEW RANGE; 2KW CONVECTOR HEATER C/W THERMOSTAT DX200	Item	24.21
DIMPLEX SKIRTING CONVECTOR; HEATER CONVECTOR SKIRTING 500W SCH5	Item	42.00

PANEL HEATERS

NEWLEC PANEL HEATERS - NEW RANGE; 1.5KW C/W TIMER NLPH1500T	Item	81.55

RADIANT FIRES

DIMPLEX; CHERITON 2KW CHT20	Item	110.63

© NSR 2008-2009

BASIC PRICES : LABOUR, PLANT AND MATERIALS

V:ELECTRICAL SERVICES : WORKS OF ALTERATION/SMALL WORKS
SPACE HEATING
ROOM THERMOSTATS

HONEYWELL; THERMOSTAT 10A 76-18308 T6360B	Item		12.93
MK ELECTRONIC ROOM THERMOSTAT; ELECTRONIC ROOM THERMOSTAT KT6360BWHI	Item		26.99
SUNVIC TLM RANGE; FROST THERMOSTAT 0.15C TLX2360	Item		21.00
SUNVIC TLM RANGE; ON/OFF 3-27C 20A TLM2402	Item		34.52

STORAGE HEATERS

DIMPLEX XLN STORAGE HEATERS; 1.7KW WHITE/GREY 718MM HX565MM WX146MM XL12N	Item		221.07

TUBULAR HEATERS

CREDA; 120W 670MM 75771502	Item		17.74
CREDA; 240W 1280MM 75771504	Item		23.70
CREDA; 360W 75771506	Item		29.40

WALL-MOUNTED FAN HEATERS

DIMPLEX - WALL MOUNTED; 3KW HIGH LEVEL MOUNTED. ELECTRONIC CONTROL FOR AUTO OPERATION CFS30	Item		147.00

WATER HEATING
HAND DRYERS

MISCELLANEOUS HAND DRIERS; HEATREA-SADIA 95.020.092 HAND DRIER 2.4KW SATIN CHROME FINISH	nr		336.00
MISCELLANEOUS HAND DRIERS; HEATSTORE HAND DRIER HEAVY DUTY AUTO 2.4KW HS4810WA	nr		347.75
MISCELLANEOUS HAND DRIERS; HEATSTORE HAND DRIER HEAVY DUTY MANUAL 2.4KW HS4800WM	nr		414.75
NEWLEC HAND DRYERS; 1.5KW HAND DRYER AUTOMATIC NLBHAA	Item		75.95
NEWLEC HAND DRYERS; 2.0KW HAND DRYER AUTOMATIC NL1AH2	Item		103.50
NEWLEC HAND DRYERS; 2.4KW AUTOMATIC HEAVY DUTY HAND + FACE DRYER NL9AHD	Item		193.50
NEWLEC HAND DRYERS; 2.4KW AUTOMATIC STAINLESS HAND DRYER NLSSHA	Item		175.95
NEWLEC HAND DRYERS; 2.4KW HAND DRYER MANUAL OPERATED NL9MHD	Item		193.50

WIRING ACCESSORIES (MOULDED & METAL FLUSH)
ASHLEY - ROCK

FUSED SPUR CONNECTION UNITS; DP SWITCH FUSED + FLEX OUTLET + NEON DSSU83N	Item		16.61

BASIC PRICES : LABOUR, PLANT AND MATERIALS

V:ELECTRICAL SERVICES : WORKS OF ALTERATION/SMALL WORKS
WIRING ACCESSORIES (MOULDED & METAL FLUSH)

ASHLEY - ROCK

FUSED SPUR CONNECTION UNITS; UN-SWITCHED + FLEX OUTLET + NEON SU83N	Item	16.16
PLATE SWITCHES - 5/6AMP; 1 GANG 1 WAY PPS11	Item	2.23
PLATE SWITCHES - 5/6AMP; 1 GANG 2 WAY PPS12	Item	2.65
PLATE SWITCHES - 5/6AMP; 1 GANG INTERMEDIATE PPS16	Item	8.09
PLATE SWITCHES - 5/6AMP; 2 GANG 2 WAY PPS22	Item	4.76
PLATE SWITCHES - 5/6AMP; 3 GANG 2 WAY PPS32	Item	6.98
PLATE SWITCHES - 5/6AMP; 4 GANG 2 WAY PPS42	Item	9.32

BG

PLATE SWITCHES 10AMP; PLATE SWITCH 6 GANG 2 WAY	Item	19.94

CRABTREE

BATTEN LAMPHOLDERS; BATTEN LAMPHOLDER - STRAIGHT 3422	Item	5.07
CEILING PULL SWITCHES; 16A 1 WAY DP 2163	Item	11.53
CEILING PULL SWITCHES; 16A 2 WAY SP 2161	Item	9.60
CEILING PULL SWITCHES; 6A 1 WAY SP 2041WH	Item	4.79
CEILING PULL SWITCHES; 6A 2 WAY SP 2141WH	Item	5.73
CEILING PULL SWITCHES; 6A RETRACTIVE SP 2147WH	Item	10.86
FUSED SPUR CONNECTION UNITS - METAL SURFACE; SWITCHED + FLEX OUTLET + NEON SATIN CHROME 1832/13SC/WH	Item	21.65
FUSED SPUR CONNECTION UNITS - METAL SURFACE; UN-SWITCHED + FLEX OUTLET SATIN CHROME 4831/3SCBK	Item	12.56
PLUG TOPS; PLUG 13A 7221WH	Item	3.18
SWITCHES - 20AMP - DP; DOUBLE POLE + NEON (MARKED WATER HEATER) 4015/31	Item	11.99
SWITCHES - 50AMP - DP (PANEL MOUNTING); 50A DOUBLE POLE + NEON (WHITE INTERIOR) 4512/13	Item	16.76

MEM

CEILING ROSES (LITELINK SYSTEM); LITELINK LSC PLUG-IN CEILING ROSE C/W 3MTRS 0.75MMSQ H/R CABLE 8572	Item	9.55
LAMPHOLDERS; LAMPHOLDER BC TYPE T2 RATING F1000	Item	1.93
PENDANT SETS; PENDANT ASSEMBLY 12 F1252	Item	6.29
PENDANT SETS; PENDANT ASSEMBLY 6 1250	Item	3.67

MK

BASIC PRICES : LABOUR, PLANT AND MATERIALS

V:ELECTRICAL SERVICES : WORKS OF ALTERATION/SMALL WORKS
WIRING ACCESSORIES (MOULDED & METAL FLUSH)

MK

CEILING PULL SWITCHES; WHITE 16A 1 WAY DP 3151	Item	8.62
CEILING PULL SWITCHES; WHITE 45A 1 WAY DP 3164	Item	13.69
CEILING PULL SWITCHES; WHITE MOUNTING BLOCK WITH NEON FOR CEILING SWITCH 2056	Item	4.02
LAMPHOLDERS; WHITE LAMPHOLDER BC PENDANT TYPE 1170	Item	2.18
LOGIC PLUS BLANK PLATES; WHITE 1 GANG BLANKPLATE K3827	Item	1.63
LOGIC PLUS BLANK PLATES; WHITE 2 GANG BLANKPLATE K3828	Item	3.65
LOGIC PLUS COOKER CONNECTION UNIT; WHITE COOKER CONNECTOR K5045	Item	6.15
LOGIC PLUS COOKER CONTROL UNITS; WHITE FLUSH METAL 45A DOUBLE POLE COOKER CONTROL + 13A OUTLET WITH NEON INDICATORS K5011	Item	35.02
LOGIC PLUS COOKER CONTROL UNITS; WHITE SURFACE METAL 45A DOUBLE POLE COOKER CONTROL + 13A OUTLET WITH NEON INDICATORS K5001	Item	55.81
LOGIC PLUS FLEX OUTLET PLATES; WHITE 1 GANG PLATE K1090	Item	5.68
LOGIC PLUS FUSED SPUR CONNECTION UNITS; WHITE 13A DP SWITCHED + FLEX OUTLET + NEON K370	Item	11.04
LOGIC PLUS FUSED SPUR CONNECTION UNITS; WHITE 13A UN-SWITCHED + FLEX OUTLET K337	Item	7.43
LOGIC PLUS RCD PROTECTED SOCKET OUTLETS; WHITE 13A 10MA RCD SWITCH SOCKET - (ACTIVE CONTROL) K6100	Item	79.03
LOGIC PLUS RCD PROTECTED SOCKET OUTLETS; WHITE 13A 30MA RCD SWITCH SOCKET - (ACTIVE CONTROL) K6300	Item	69.36
LOGIC PLUS SOCKET OUTLETS - UNSWITCHED; WHITE 13A 2 GANG SOCKET OUTLET FLUSH K781	Item	10.03
LOGIC PLUS SOCKET OUTLETS - UNSWITCHED; WHITE 15A 3 PIN SOCKET OUTLET FLUSH K772	Item	12.97
LOGIC PLUS SWITCH AND SOCKET BOXES - SURFACE STEEL; 1 GANG STEEL SURFACE BOX 41MM K2211 ALM	Item	3.79
LOGIC PLUS SWITCH AND SOCKET BOXES - SURFACE STEEL; 2 GANG STEEL SURFACE BOX 41MM WITH KNOCKOUTS K2214 ALM	Item	5.13
LOGIC PLUS SWITCH AND SOCKET MOUNTING BOXES (MOULDED); WHITE 1 GANG BOX SURFACE 44MM DEEP K2031	Item	2.07
LOGIC PLUS SWITCH AND SOCKET MOUNTING BOXES (MOULDED); WHITE 1 GANG SURFACE BOX 32MM DEEP K2140	Item	1.56
LOGIC PLUS SWITCH AND SOCKET MOUNTING BOXES (MOULDED); WHITE 2 GANG FLANGE BOX 45MM DEEP K2062	Item	3.23

BASIC PRICES : LABOUR, PLANT AND MATERIALS

V:ELECTRICAL SERVICES : WORKS OF ALTERATION/SMALL WORKS
WIRING ACCESSORIES (MOULDED & METAL FLUSH)

MK

Description	Unit	Price
LOGIC PLUS SWITCH AND SOCKET MOUNTING BOXES (MOULDED); WHITE 2 GANG SURFACE BOX 32MM DEEP K2142	Item	2.96
LOGIC PLUS SWITCHED SOCKET OUTLETS; WHITE 13A 1 GANG + NEON K2657	Item	10.00
LOGIC PLUS SWITCHED SOCKET OUTLETS; WHITE 13A 1 GANG K2757	Item	3.71
LOGIC PLUS SWITCHED SOCKET OUTLETS; WHITE 13A 2 GANG + NEONS K2647	Item	15.50
LOGIC PLUS SWITCHED SOCKET OUTLETS; WHITE 13A 2 GANG K2747	Item	6.83
LOGIC PLUS SWITCHED SOCKET OUTLETS; WHITE 15A 1 GANG K2893	Item	16.56
LOGIC PLUS SWITCHES - 20 AMP - DP; WHITE 20A DP + FLEX-OUTLET WITH NEON INDICATOR AND RED ROCKER K5423D1WHI	Item	15.90
LOGIC PLUS SWITCHES - 32 AMP - DP; WHITE 32A DOUBLE POLE SWITCH WITH NEON K5105	Item	17.27
LOGIC PLUS SWITCHES - 45 AMP - DP; WHITE 45A WITH NEON K5215	Item	16.28
PLUG TOPS; ROUND PIN 2A 3 PIN 502WHITE	Item	3.08
RCD PROTECTED SOCKET OUTLETS - ALBANY/CHROMA PLUS; 13A SINGLE RCD SOCKET 30MA ACTIVE - MATT CHROME K6301 MCO	Item	94.54
SOCKET OUTLETS - UNSWITCHED; WHITE 2A 3 PIN CIRCULAR (BESA BOX FIXING) 312	Item	7.46
SWITCH SOCKET OUTLETS - EDGE RANGE - NEW RANGE; 13A 1 GANG DP SILVER ANODISED ALUMINIUM DUAL EARTH C/W WHITE INSERTS K14357 SAA	Item	17.16
SWITCH SOCKET OUTLETS - EDGE RANGE - NEW RANGE; 13A 2 GANG DP SILVER ANODISED ALUMINIUM DUAL EARTH C/W WHITE INSERTS K14347 SAA	Item	30.20

THORPE

Description	Unit	Price
LAMPHOLDERS - PORCELAIN; ES PLAIN CORD GRIP ENTRY LH5237	Item	7.27

WIRING ACCESSORIES SURFACE METAL CLAD

ASHLEY - ROCK

Description	Unit	Price
FUSED SPUR CONNECTION UNITS - METALCLAD; SWITCHED + FLEX-OUTLET MSSU3	Item	14.99
FUSED SPUR CONNECTION UNITS - METALCLAD; SWITCHED + NEON + FLEX-OUTLET MSSU3N	Item	19.31
SWITCH - METALCLAD 20AMP; 20A DOUBLE POLE + FLEX OUTLET 1 GANG MDP4F	Item	18.54

MK

BASIC PRICES : LABOUR, PLANT AND MATERIALS

V:ELECTRICAL SERVICES : WORKS OF ALTERATION/SMALL WORKS
WIRING ACCESSORIES SURFACE METAL CLAD

MK

Description	Unit	Price
FUSED SPUR CONNECTION UNITS - METALCLAD PLUS; 13A 1 GANG RCD PROTECTED SWITCH SOCKET 30MA GREY 56301	Item	80.69
FUSED SPUR CONNECTION UNITS - METALCLAD PLUS; 13A DOUBLE POLE SWITCHED + FLEX OUTLET K989 ALM	Item	11.93
FUSED SPUR CONNECTION UNITS - METALCLAD PLUS; 13A DOUBLE POLE SWITCHED + NEON + FLEX OUTLET K972 ALM	Item	14.20
SWITCHED SOCKET OUTLETS - METALCLAD PLUS; 13A 1 GANG DOUBLE POLE + NEON K2477 ALM	Item	20.44
SWITCHED SOCKET OUTLETS - METALCLAD PLUS; 13A 1 GANG DOUBLE POLE K2977 ALM	Item	9.89
SWITCHED SOCKET OUTLETS - METALCLAD PLUS; 13A 2 GANG DOUBLE POLE K2946 ALM	Item	15.56
SWITCHED SOCKET OUTLETS - METALCLAD PLUS; 13A 2 GANG DOUBLE POLE + NEON K2446 ALM	Item	34.05

BASIC PRICES : LABOUR, PLANT AND MATERIALS

W:ELECTRICAL SERVICES : WORKS OF ALTERATION/SMALL WORKS
BATTERIES & TORCHES
BATTERY CHARGERS

YUASA; 24VOLT 4AMP BATTERY CHARGER 3STAGE SWITCHMODE TYPE C4A24		Item	110.00

CABLE & CABLE ACCESSORIES
FLEXIBLE CORDS

2183Y PVC INSULATED/PVC SHEATHED - 3 CORE CIRCULAR; 100MTR 0.5MM		1000m	952.30

PYROTENAX MINERAL INSULATED CABLES AND ACCESSORIES

COMPRESSION GLANDS - BRASS (HEAVY DUTY); COMPRESSION GLAND FOR 2H2.5 CABLE RGM2H2.5		Item	32.42
COMPRESSION GLANDS - BRASS (HEAVY DUTY); COMPRESSION GLAND FOR 2H4 CABLE RGM2H4		Item	32.42
COMPRESSION GLANDS - BRASS (HEAVY DUTY); COMPRESSION GLAND FOR 2H6 CABLE RGM2H6		Item	32.42
COMPRESSION GLANDS - BRASS (HEAVY DUTY); COMPRESSION GLAND FOR 3H1.5 CABLE RGM3H1.5		Item	32.42
COMPRESSION GLANDS - BRASS (HEAVY DUTY); COMPRESSION GLAND FOR 3H2.5 CABLE RGM3H2.5		Item	32.42
COMPRESSION GLANDS - BRASS (HEAVY DUTY); COMPRESSION GLAND FOR 3H4 CABLE RGM3H4		Item	32.42
COMPRESSION GLANDS - BRASS (HEAVY DUTY); COMPRESSION GLAND FOR 3H6 CABLE RGM3H6		Item	21.60
COMPRESSION GLANDS - BRASS (HEAVY DUTY); COMPRESSION GLAND FOR 4H1.5 CABLE RGM4H1.5		Item	32.42
COMPRESSION GLANDS - BRASS (HEAVY DUTY); COMPRESSION GLAND FOR 4H4 CABLE RGM4H4		Item	21.60
COMPRESSION GLANDS - BRASS (HEAVY DUTY); COMPRESSION GLAND FOR 4H4 CABLE RGM4H4		Item	21.60
COMPRESSION GLANDS - BRASS (HEAVY DUTY); COMPRESSION GLAND FOR 4H6 CABLE RGM4H6		Item	21.60
COMPRESSION GLANDS - BRASS (LIGHT DUTY); COMPRESSION GLAND FOR 2L1.5 CABLE RGM2L1.5		Item	6.36
COMPRESSION GLANDS - BRASS (LIGHT DUTY); COMPRESSION GLAND FOR 2L2.5 CABLE RGM2L2.5		Item	6.36
COMPRESSION GLANDS - BRASS (LIGHT DUTY); COMPRESSION GLAND FOR 3L1.5 CABLE RGM3L1.5		Item	6.36
COMPRESSION GLANDS - BRASS (LIGHT DUTY); COMPRESSION GLAND FOR 4L1.5 CABLE RGM4L1.5		Item	6.36
COMPRESSION GLANDS - BRASS (LIGHT DUTY); COMPRESSION GLAND FOR 4L2.5 CABLE RGM4L2.5		Item	6.36
GLAND SHROUDS - PVC AND LSF TYPES; RED GLAND SHROUD LSF 25MM RHGMM25		Item	19.63
GLAND SHROUDS - PVC AND LSF TYPES; RED GLAND SHROUD PVC 20MM RHG20		Item	10.90
GLAND SHROUDS - PVC AND LSF TYPES; RED LSF 25MM RHGMM25		Item	19.63

BASIC PRICES : LABOUR, PLANT AND MATERIALS

W:ELECTRICAL SERVICES : WORKS OF ALTERATION/SMALL WORKS
CABLE & CABLE ACCESSORIES
PYROTENAX MINERAL INSULATED CABLES AND ACCESSORIES

Description	Unit	Price
GLAND SHROUDS - PVC AND LSF TYPES; RED PVC 20MM RHG20	Item	10.90
GLAND SHROUDS - PVC AND LSF TYPES; WHITE GLAND SHROUD PVC 20MM RHG20	Item	10.90
LIGHT DUTY - BARE COPPER SHEATHED - 500V GRADE - 100MTR DRUM; 1.5MM 2CORE CC2L1.5	100m	693.30
LIGHT DUTY - BARE COPPER SHEATHED - 500V GRADE - 100MTR DRUM; 1.5MM 2CORE CC2L1.5	100m	693.30
LIGHT DUTY - BARE COPPER SHEATHED - 500V GRADE - 100MTR DRUM; 1.5MM 3CORE CC3L1.5	100m	917.20
LIGHT DUTY - BARE COPPER SHEATHED - 500V GRADE - 100MTR DRUM; 1.5MM 4CORE CC4L1.5	100m	1,077.40
LIGHT DUTY - BARE COPPER SHEATHED - 500V GRADE - 100MTR DRUM; 1.5MM 7CORE CC7L1.5	100m	1,664.10
LIGHT DUTY - BARE COPPER SHEATHED - 500V GRADE - 100MTR DRUM; 2.5MM 2CORE CC2L2.5	100m	900.60
LIGHT DUTY - LSF SHEATHED BLACK - 500 GRADE - 100MTR DRUM; 1.5MM 4CORE CCM4L1.5	100m	1,202.50
LIGHT DUTY - LSF SHEATHED BLACK - 500 GRADE - 100MTR DRUM; 2.5MM 2CORE CCM2L2.5	100m	961.70
LIGHT DUTY - LSF SHEATHED ORANGE - 500V GRADE - (CUT TO LENGTH SERVICE); 2.5MM 3CORE CCM3L2.5	m	15.20
LIGHT DUTY - LSF SHEATHED ORANGE - 500V GRADE - (CUT TO LENGTH SERVICE); 2.5MM 4CORE CCM4L2.5	m	18.45
LIGHT DUTY - LSF SHEATHED ORANGE - 500V GRADE - (CUT TO LENGTH SERVICE); 4.0MM 2CORE CCM2L4	m	14.47
LIGHT DUTY - LSF SHEATHED ORANGE - 500V GRADE - 100MTR DRUM; 1.0MM 2CORE CCM2L1	100m	686.20
LIGHT DUTY - LSF SHEATHED ORANGE - 500V GRADE - 100MTR DRUM; 1.0MM 3CORE CCM3L1	100m	832.50
LIGHT DUTY - LSF SHEATHED ORANGE - 500V GRADE - 100MTR DRUM; 1.5MM 2CORE CCM2L1.5	100m	761.60
LIGHT DUTY - LSF SHEATHED ORANGE - 500V GRADE - 100MTR DRUM; 1.5MM 2CORE 100MTR 500V LSF 2L1.5	100m	761.60
LIGHT DUTY - LSF SHEATHED ORANGE - 500V GRADE - 100MTR DRUM; 1.5MM 3CORE 500V LSF 3L1.5	100m	1,015.90
LIGHT DUTY - LSF SHEATHED ORANGE - 500V GRADE - 100MTR DRUM; 1.5MM 3CORE CCM3L1.5	100m	1,015.90
LIGHT DUTY - LSF SHEATHED ORANGE - 500V GRADE - 100MTR DRUM; 1.5MM 4CORE CCM4L1.5	100m	1,202.50
LIGHT DUTY - LSF SHEATHED ORANGE - 500V GRADE - 100MTR DRUM; 2.5MM 2CORE CCM2L2.5	m	9.62
LIGHT DUTY - LSF SHEATHED ORANGE - 500V GRADE - 100MTR DRUM; 2.5MM 3CORE CCM3L2.5	100m	1,520.30
LIGHT DUTY - LSF SHEATHED RED - 500V GRADE - (CUT TO LENGTH SERVICE); CUTS 4.0MM 2CORE CCM2L4	m	14.47

BASIC PRICES : LABOUR, PLANT AND MATERIALS

W:ELECTRICAL SERVICES : WORKS OF ALTERATION/SMALL WORKS
CABLE & CABLE ACCESSORIES

PYROTENAX MINERAL INSULATED CABLES AND ACCESSORIES

Description	Unit	Price
LIGHT DUTY - LSF SHEATHED RED - 500V GRADE - 100MTR DRUM; 1.0MM 4CORE 100MTR CCM4L1	100m	978.40
LIGHT DUTY - LSF SHEATHED WHITE - 500V GRADE - (CUT TO LENGTH SERVICE); CUTS 2.5MM 4CORE CCM4L2.5	m	18.15
LIGHT DUTY - LSF SHEATHED WHITE - 500V GRADE - (CUT TO LENGTH SERVICE); CUTS 4.0MM 2CORE CCM2L4	m	14.47
SEAL ASSEMBLIES - (HEAVY DUTY) EARTH TAIL TYPE; 2H6 CABLE RPSL2H6	Item	27.90
SEAL ASSEMBLIES - (HEAVY DUTY) STANDARD SCREW-ON TYPE; SEAL SCREW-ON FOR 2H4 CABLE RPS2H4	Item	21.15
SEAL ASSEMBLIES - (HEAVY DUTY) STANDARD SCREW-ON TYPE; SEAL SCREW-ON FOR 2H4 CABLE RPS2H4	Item	21.15
SEAL ASSEMBLIES - (HEAVY DUTY) STANDARD SCREW-ON TYPE; SEAL SCREW-ON FOR 2H6 CABLE RPS2H6	Item	21.15
SEAL ASSEMBLIES - (HEAVY DUTY) STANDARD SCREW-ON TYPE; SEAL SCREW-ON FOR 3H1.5 CABLE RPS3H1.5	Item	21.54
SEAL ASSEMBLIES - (HEAVY DUTY) STANDARD SCREW-ON TYPE; SEAL SCREW-ON FOR 3H2.5 CABLE RPS3H2.5	Item	21.54
SEAL ASSEMBLIES - (HEAVY DUTY) STANDARD SCREW-ON TYPE; SEAL SCREW-ON FOR 3H4 CABLE RPS3H4	Item	21.54
SEAL ASSEMBLIES - (HEAVY DUTY) STANDARD SCREW-ON TYPE; SEAL SCREW-ON FOR 3H6 CABLE RPS3H6	Item	19.97
SEAL ASSEMBLIES - (HEAVY DUTY) STANDARD SCREW-ON TYPE; SEAL SCREW-ON FOR 4H1.5 CABLE RPS4H1.5	Item	21.54
SEAL ASSEMBLIES - (HEAVY DUTY) STANDARD SCREW-ON TYPE; SEAL SCREW-ON FOR 4H2.5 CABLE RPS4H2.5	Item	21.54
SEAL ASSEMBLIES - (HEAVY DUTY) STANDARD SCREW-ON TYPE; SEAL SCREW-ON FOR 4H4 CABLE RPS4H4	Item	19.97
SEAL ASSEMBLIES - (HEAVY DUTY) STANDARD SCREW-ON TYPE; SEAL SCREW-ON FOR 4H6 CABLE RPS4H6	Item	19.97
SEAL ASSEMBLIES - (LIGHT DUTY) EARTH TAIL TYPE; EARTH TAIL SEAL FOR 3L2.5 CABLE RPSL3L2.5	Item	40.46
SEAL ASSEMBLIES - (LIGHT DUTY) STANDARD SCREW-ON TYPE; SEAL SCREW-ON FOR 2L1.5 CABLE RPS2L1.5	Item	20.79
SEAL ASSEMBLIES - (LIGHT DUTY) STANDARD SCREW-ON TYPE; SEAL SCREW-ON FOR 2L2.5 CABLE RPS2L2.5	Item	20.79

© NSR 2008-2009

BASIC PRICES : LABOUR, PLANT AND MATERIALS

W:ELECTRICAL SERVICES : WORKS OF ALTERATION/SMALL WORKS
CABLE & CABLE ACCESSORIES
PYROTENAX MINERAL INSULATED CABLES AND ACCESSORIES

Description	Unit	Price
SEAL ASSEMBLIES - (LIGHT DUTY) STANDARD SCREW-ON TYPE; SEAL SCREW-ON FOR 3L1.5 CABLE RPS3L1.5	Item	21.54
SEAL ASSEMBLIES - (LIGHT DUTY) STANDARD SCREW-ON TYPE; SEAL SCREW-ON FOR 4L1.5 CABLE RPS4L1.5	Item	21.54
SEAL ASSEMBLIES - (LIGHT DUTY) STANDARD SCREW-ON TYPE; SEAL SCREW-ON FOR 4L2.5 CABLE RPS4L2.5	Item	21.54

CABLE MANAGEMENT SYSTEMS
FLEXIBLE CONDUIT (NON-METALLIC) AND FITTINGS

Description	Unit	Price
ADAPTAFLEX XTRAFLEX PVC CONDUIT AND FITTINGS; 20MM ADAPTOR XF20M20D	Item	2.27
ADAPTAFLEX XTRAFLEX PVC CONDUIT AND FITTINGS; 20MM PVC CONDUIT 30MTR COIL XF20	1000m	93.86
ADAPTAFLEX XTRAFLEX PVC CONDUIT AND FITTINGS; 25MM ADAPTOR XF25M25D	Item	2.64
ADAPTAFLEX XTRAFLEX PVC CONDUIT AND FITTINGS; 25MM PVC CONDUIT 30MTR COIL XF25	Item	116.45

FLOORBOX SYSTEMS

Description	Unit	Price
NEWLEC FLOORBOX AND ACCESSORIES; CAVITY FLOORBOX 3 COMPARTMENT NLFB2N	Item	21.95
NEWLEC FLOORBOX AND ACCESSORIES; OUTLET PANEL ACCEPTS 2 X LJ2 TYPE ACCESSORIES NLFB2P2N	Item	2.00
NEWLEC FLOORBOX AND ACCESSORIES; OUTLET PANEL FITTED WITH 1 X TWIN SWITCHED SOCKET (DUAL EARTH) NLFB2P5EN	Item	7.50

TRUNKING AND FITTINGS PLASTIC

Description	Unit	Price
MARSHALL-TUFFLEX SCEPTRE DADO TRUNKING; CONNECTOR 100X25MM DCJ2	Item	6.57
MARSHALL-TUFFLEX SCEPTRE DADO TRUNKING; DADO TRUNKING 100X25MM 3MTR LENGTH DTR2	Item	84.76
MARSHALL-TUFFLEX SCEPTRE DADO TRUNKING; DIVIDING FILLET 100X25MM 3MTR LENGTH DDF2	Item	8.75
MITA CABLELINE CLASSIC SKIRTING & DADO TRUNKING - WHITE; CABLELINE LID (3MTR LENGTH) CSN3W	Item	18.15
MITA CABLELINE CLASSIC SKIRTING & DADO TRUNKING - WHITE; CABLELINE SOCKET BOX 1 GANG (VERTICAL) 25MM CLB3-1/4W	Item	2.76
MITA CABLELINE CLASSIC SKIRTING & DADO TRUNKING - WHITE; CLASSIC TRUNKING 145X40MM 3MTR LENGTH CS3W	Item	75.54
MITA CABLELINE CLASSIC SKIRTING & DADO TRUNKING - WHITE; JOINT COVER TO SUIT CS3 PROFILE CSJ3W	Item	3.20

BASIC PRICES : LABOUR, PLANT AND MATERIALS

W:ELECTRICAL SERVICES : WORKS OF ALTERATION/SMALL WORKS

CABLE MANAGEMENT SYSTEMS

TRUNKING AND FITTINGS PLASTIC

MITA CONSORT DADO TRUNKING - WHITE; CABLE RETAINER 100X40MM CTR104	Item	0.77
MITA CONSORT DADO TRUNKING - WHITE; CABLE RETAINER 100X60MM CTR106W	Item	0.93
MITA CONSORT DADO TRUNKING - WHITE; COUPLER 100X40MM (PAIR) CTC40W	Item	0.91
MITA CONSORT DADO TRUNKING - WHITE; DADO TRUNKING 100X40MM 3MTR LENGTH CT104W	Item	36.93
MITA CONSORT DADO TRUNKING - WHITE; DADO TRUNKING 100X60MM 3MTR LENGTH CT106W	Item	51.72
MITA CONSORT DADO TRUNKING - WHITE; DADO TRUNKING 130X60MM 3MTR LENGTH CT136W	Item	62.31
MITA CONSORT DADO TRUNKING - WHITE; JOINT COVER 100X40MM CTJ104	Item	2.39
MITA CONSORT DADO TRUNKING - WHITE; JOINT COVER 100X60MM CTJ106W	Item	2.85
MITA CONSORT DADO TRUNKING - WHITE; JOINT COVER 130X60MM CTJ136W	Item	3.45
MK PREMIER COMMERCIAL DADO TRUNKING - WHITE; 100MMX100MM 3MTR LENGTH NCT1010	Item	97.95
MK PREMIER COMMERCIAL DADO TRUNKING - WHITE; 100MMX50MM 3MTR LENGTH NCT1050	Item	78.72

FIRE ALARMS & SECURITY SYSTEMS

ACCESS CONTROL SYSTEMS

DOOR CONTROLS; CARD READER AC-BC402	Item	172.99
DOOR CONTROLS; CARD READER AC-BC402	Item	172.99
DOOR CONTROLS; ELECTRIC MORTICE STRIKE NLLRM3	Item	17.95
DOOR CONTROLS; ELECTRIC RIM STRIKE NLLRR2	Item	14.95
DOOR CONTROLS; ELECTRONIC DIGITAL CODELOCK POLISHED BRASS 5020	Item	215.00

BELLS & SOUNDERS

BELLS & SOUNDERS; 12V DC GREY MASTERBELL 66-12SP	Item	21.26
BELLS & SOUNDERS; 240V AC GREY BELL MMB240	Item	34.50
BELLS & SOUNDERS; 24V DC RED BELL MASTERBELL 66-24SPR	Item	21.26
BELLS & SOUNDERS; BELL 4 24V INTERNAL TYPE MBM244	Item	19.60
BELLS & SOUNDERS; BELL 8 24V INTERNAL TYPE MBM248	Item	30.00
BELLS & SOUNDERS; C-TEC 24V SIREN	Item	18.85

BREAK GLASS CALL POINTS

BASIC PRICES : LABOUR, PLANT AND MATERIALS

W:ELECTRICAL SERVICES : WORKS OF ALTERATION/SMALL WORKS
FIRE ALARMS & SECURITY SYSTEMS

BREAK GLASS CALL POINTS

CHLORIDE CALL POINTS - ZIRCON RANGE; BREAK GLASS CALL POINT SURFACE MOUNTED ZF50	Item	8.81
CHLORIDE CALL POINTS - ZIRCON RANGE; WEATHERPROOF BREAK GLASS CALL POINT - IP67 ZF55	Item	27.47
EMERGELITE CALL POINT; CALL POINT SURFACE MOUNTED (NO LED) FMC-800	Item	10.70
EMERGELITE CALL POINT; REPLACEMENT SPARE GLASS (PACK 10) FMC-10G	Item	13.00

CARBON MONOXIDE DETECTORS

ALTO CARBON MONOXIDE ALARMS; CARBON MONOXIDE ALARM BATTERY POWERED AL4304	Item	28.05

CLOSED CIRCUIT TELEVISION SYSTEMS (CCTV)

SECURITY CCTV SYSTEMS; 12 MONOCHROME MONITOR INC METAL CASE CCTV-CEM1211	Item	86.78
SECURITY CCTV SYSTEMS; 14 COLOUR MONITOR INC METAL CASE CCTV-ZMCR114NP	Item	175.05
SECURITY CCTV SYSTEMS; 90° BLACK & WHITE CAMERA C/W 20MTR CABLE AL9162	Item	89.05
SECURITY CCTV SYSTEMS; CAMERA COVERT CCD MONOCHROME PIR AL2662	Item	125.60

FIRE ALARM CONTROL PANELS

MENVIER FIRE PANEL - 9300 RANGE; 12-16 ZONE FIRE ALARM PANEL C/W BATTERY EN54 MF9316	Item	737.90
MENVIER FIRE PANEL - 9300 RANGE; 2 ZONE FIRE ALARM PANEL MF9302	Item	221.90
MENVIER FIRE PANEL - 9300 RANGE; 6-8 ZONE FIRE ALARM PANEL (24HR STANDBY) MF9308L	Item	431.35
MENVIER FIRE PANEL - 9300 RANGE; 8 ZONE FIRE ALARM PANEL MF9308	Item	527.95
RECHARGEABLE BATTERIES (FOR FIRE ALARM / BURGLAR ALARM CONTROL PANELS); 12V 17AH BATTERY NP15-12	Item	81.67

SMOKE AND HEAT DETECTORS

AICO SMOKE ALARM FOR THE DEAF & HARD OF HEARING; DEAF ALARM KIT FOR 150 SERIES (INCL. POWER PACK WITH BUILT-IN STROBE + VIBRATING DISC) _EI169	Item	130.00
AICO SMOKE ALARMS - 160 SERIES - NEW RANGE; IONISATION SMOKE ALARM C/W HUSH BUTTON + LITHIUM CELL + MOUNTING PLATE EI161	Item	26.51
AICO SMOKE ALARMS - 160 SERIES - NEW RANGE; OPTICAL SMOKE ALARM C/W HUSH BUTTON + LITHIUM CELL + MOUNTING PLATE EI166	Item	37.42
AICO SMOKE ALARMS - MAINS POWERED - 150 SERIES; IONISATION SMOKE ALARM C/W HUSH BUTTON + BUILT-IN RECHARGEABLE LITHIUM CELLS EI151TL	Item	27.25

BASIC PRICES : LABOUR, PLANT AND MATERIALS

W:ELECTRICAL SERVICES : WORKS OF ALTERATION/SMALL WORKS

FIRE ALARMS & SECURITY SYSTEMS

SMOKE AND HEAT DETECTORS

CHLORIDE DETECTORS - ZIRCON RANGE; FIXED HIGH TEMPERATURE HEAT DETECTOR 60° C ZF62	Item	14.25
HEAT ALARM - MAINS POWERED; MAINS HEAT ALARM C/W ALKALINE BATTERY BACK-UP AL4303	Item	27.25
NEWLEC DOMESTIC SMOKE ALARM; MAINS OPERATED WITH BATTERY BACK UP 240V AC IONISATION TYPE NL3602	Item	9.75

GENERAL

Sundries

Materials; Unit material rate	BASE	1.35